James II and the Trial of the Seven Bishops

Also by William Gibson

RELIGION AND THE ENLIGHTENMENT, 1600–1800: Conflict and the Rise of Civic Humanism in Taunton

THE ENLIGHTENMENT PRELATE; BENJAMIN HOADLY 1676–1761

THE CHURCH OF ENGLAND 1688–1832: Unity and Accord

RELIGION AND SOCIETY IN ENGLAND AND WALES, 1689–1800

A SOCIAL HISTORY OF THE DOMESTIC CHAPLAIN, 1530–1840

THE ACHIEVEMENT OF THE ANGLICAN CHURCH, 1689–1800: The Confessional State in England in the Eighteenth Century Church, State and Society, 1760–1850

James II and the Trial of the Seven Bishops

William Gibson
Professor of Ecclesiastical History,
Oxford Brookes University

© William Gibson 2009

All rights reserved. No reproduction, copy or transmission of this publication may be made without written permission.

No portion of this publication may be reproduced, copied or transmitted save with written permission or in accordance with the provisions of the Copyright, Designs and Patents Act 1988, or under the terms of any licence permitting limited copying issued by the Copyright Licensing Agency, Saffron House, 6-10 Kirby Street, London EC1N 8TS.

Any person who does any unauthorized act in relation to this publication may be liable to criminal prosecution and civil claims for damages.

The author has asserted his right to be identified as the author of this work in accordance with the Copyright, Designs and Patents Act 1988.

First published 2009 by
PALGRAVE MACMILLAN

Palgrave Macmillan in the UK is an imprint of Macmillan Publishers Limited, registered in England, company number 785998, of Houndmills, Basingstoke, Hampshire RG21 6XS.

Palgrave Macmillan in the US is a division of St Martin's Press LLC,
175 Fifth Avenue, New York, NY 10010.

Palgrave Macmillan is the global academic imprint of the above companies and has companies and representatives throughout the world.

Palgrave® and Macmillan® are registered trademarks in the United States, the United Kingdom, Europe and other countries.

ISBN-13: 978–0–230–20400–3 hardback
ISBN-10: 0–230–20400–7 hardback

This book is printed on paper suitable for recycling and made from fully managed and sustained forest sources. Logging, pulping and manufacturing processes are expected to conform to the environmental regulations of the country of origin.

A catalogue record for this book is available from the British Library.

Library of Congress Cataloging-in-Publication Data

Gibson, William, 1959–
  James II and the trial of the seven bishops / William Gibson.
    p. cm.
  Includes bibliographical references and index.
    ISBN-13: 978–0–230–20400–3 ISBN-10: 0–230–20400–7
    1. Great Britain – Politics and government – 1660–1688. 2. James II, King of England, 1633–1701. 3. Church of England – Bishops – History – 17th century. 4. Trials (Seditious libel) – England. 5. Great Britain – History – Revolution of 1688. 6. Great Britain – History – James II, 1685–1688. I. Title.

DA452.G53 2009
941.06'7—dc22   2008037194

10 9 8 7 6 5 4 3 2 1
18 17 16 15 14 13 12 11 10 09

Printed and bound in Great Britain by
CPI Antony Rowe, Chippenham and Eastbourne

For John Walsh and John Morgan-Guy

Contents

Preface	ix

Introduction: The Seven Bishops and the Glorious Revolution		1
Echoes of the trial		1
Macaulay's heroes and villains		3
James and revisionism		6
James's rehabilitation		10
The new Whig Revolution		12
Tory views atomised		14
Continuity and change		16
The seven bishops		17
1	**The Bishops: Unlikely Revolutionaries**	**20**
	Family and background	20
	Education and training	21
	Early preferment	24
	Consecrations	29
	Preoccupations	34
	Sceptical loyalism	44
2	**The King's Policies 1685–7**	**47**
	Early days and Monmouth's rebellion	47
	Compton and the Ecclesiastical Commission	53
	The attack on the universities	62
	The first declaration	64
	The attack on the landed alliance	68
3	**The Confrontation**	**73**
	The gathering storm	73
	The second declaration	78
	Canvassing opposition	83
	Petitioning the King	85
	The leak of the petition	90
	The propaganda war	92

viii *Contents*

4	**The Tower**	**97**
	Silent pulpits	97
	The legal case	98
	The empty country pulpits	100
	The second confrontation	105
	To the tower	109
5	**The Trial**	**114**
	The bishops arraigned	114
	The bishops released	117
	The 'bed-pan baby'	121
	The trial in the King's Bench	123
	The acquittal	132
6	**The Reaction**	**139**
	The stiffened resolve	139
	The pursuit of the clergy	144
	The wooing of dissent	150
	The Anglican revolution in the state	152
	Preparations for the invasion	159
7	**The Revolution**	**162**
	The invasion	162
	James at bay	167
	The Guildhall meeting	172
	The convention	175
	Sancroft's departure	180
	Schism and expulsions	183
Conclusion		**194**
Notes		204
Bibliography		231
Index		243

Preface

When the Palace of Westminster was built by Sir Charles Barry, following the fire of 1834, an important place in the Commons North Corridor was found for a fresco of the *Acquittal of the Seven Bishops* by E. M. Ward. In it, Archbishop Sancroft, leaning heavily on a walking stick, leads the seven bishops from Westminster Hall, giving their blessings to the assembled people who kneel before them. The image was regarded as a moment in British history which deserved a place in Parliament, alongside other great occasions in history, such as King Alfred raising the Saxons against the Danes; King John granting Magna Carta; Speaker Lenthall asserting the Commons' privileges against Charles I; Henry VII chartering Cabot; the Commons petitioning Queen Elizabeth I; and the embarkation of the pilgrim fathers. It was one of those pivotal moments in British history and its depiction was intended to remind peers and MPs that leaders might be brought to heel by the rule of law.

The Glorious Revolution in 1688 has undergone something of a revival of late; six recent studies have analysed the Revolution from one perspective or another.[1] While the explicit Whig accounts of the Glorious Revolution as the start of Britain's inexorable rise to progress and liberty are long gone, a distinct whiff of Whiggery remains in some accounts. Nevertheless what few of these books do is to place religion, and particularly the role of the petition, imprisonment, trial and acquittal of the seven bishops in their true perspective as central to the events of 1688. In this, at least, the Whig historians saw the Glorious Revolution aright. G. M. Trevelyan commented that the trial of the bishops was

> the greatest historical drama that ever took place before an authorised English law court...it showed as nothing else would have done that the most revered and the most loyal subjects in the land would be broken if they refused to become active parties to the King's illegal designs. If the bishops suffered, who could hope to escape the royal vengeance?[2]

Seen from any angle, what happened in England in 1688–9 is that men and women chose Protestantism over monarchy – perhaps the last occasion on which the commitment of people to the Church exceeded that to the crown. Even James II (and VII of Scotland), after the Revolution, ascribed his loss of the throne to the 'groundless' fears the Church had about Catholicism.[3] James's reckless decision to prosecute the bishops was, in the words of G. M. Trevelyan, to throw 'into the loaded mine the lighted match of the trial of the seven bishops.'[4] But it was testimony to James's failure to understand the loyalty to the Church that most people felt that he behaved in such an incendiary fashion. Indeed, J. R. Jones called the bishops' petition 'a national gesture of resistance' to James.[5]

This study grew out of my work on Taunton in this period. Taunton exemplified James's miscalculation: with a large Dissenter population, it was one of those towns which James might have expected to welcome the Declarations of Indulgence in 1687 and 1688. But Taunton was also a centre of stern and unbending Protestantism, few would throw in their lot with a Catholic king against their Protestant, albeit persecuting, Anglican brothers.

In the course of writing this book, I have incurred a great many debts. I owe a debt of gratitude to those librarians and archivists who patiently answered my questions and aided my access to books and manuscripts. I am pleased to acknowledge the research fund of Westminster Institute of Education for supporting my visits to the Huntington Library in California to work on the Ellesmere, Hastings and Huntingdon papers and to the National Library of Wales. I am grateful to Professor Jeremy Black of Exeter University, Dr Robert Ingram of Ohio University, Dr Newton Key of Eastern Illinois University, Dr John Morgan-Guy of University of Wales Lampeter and Mary Connolly for reading an earlier draft of the manuscript, and saving me from many errors. I am indebted to Michael Strang of Palgrave Macmillan for being interested enough in this idea to publish it. This book is dedicated to John Walsh and John Morgan-Guy, two friends and colleagues whose conversation on this topic was a source of ideas and encouragement throughout the writing of this book.

Introduction: The Seven Bishops and the Glorious Revolution

Echoes of the trial

On 16 October 1706, a hackney coachman of Coal Yard near Drury Lane reported that he had heard John Baldwin assert that the Pretender *was* the true son of James II. When he was challenged on James's imprisonment of the seven bishops in the Tower, Baldwin replied 'Damn the Bishops'.[1] Thus eighteen years after they had been imprisoned and tried, the seven bishops, and their role in the overthrow of James II, were alive in the minds and lore of the working people of London. For men and women of the early eighteenth century, Catholicism and France represented twin evils, from which the events of 1688 had saved them. Only a year before Baldwin's misbehaviour, a broadsheet entitled *Great and Good News to the Church of England* calculated that Anglicans outnumbered Catholics and Dissenters by 102 to 1, and included the text of the seven bishops' speech to James when they presented their petition to him. The broadsheet also pointed out that the days on which the bishops had been imprisoned and released had particularly appropriate – and therefore apparently providential – readings from the Book of Common Prayer.[2]

Five years later, after Henry Sacheverell had been tried before the House of Lords for sedition in claiming that Low Churchmen were 'false brethren' and questioning the legitimacy of the Glorious Revolution, the actions of the seven bishops were also recalled. The anonymous author of *A History of King James's Ecclesiastical Commission*, published in 1711, claimed that while the people of England had been 'very sensible' of James's designs, it was 'the clergy [who] in a more particular manner thought it a duty high incumbent on them not to be silent in the common danger'. This vigilance of the clergy to political danger was a common theme in 1710–11. Bishops Crewe of Durham and Sprat of Rochester both lived long enough to have sat on James's Ecclesiastical Commission and survive to 1711; their support for James meant 'we may make a judgement of the zeal they signalized in the case of Sacheverell for whom they voted as heartily as they had given their votes against the Bishop of London'.[3]

For years to come, the trial of the seven bishops was used to good effect when defenders of the Church wanted to provide examples of clerical suffering. In the period between 1710 and 1714, as High Churchmen dominated the convocation of the Church and sought to suppress Low Churchmen, some saw parallels with 1688. A tract entitled *What Has Been May Be: Or a View of a Popish and an Arbitrary Government ... To which Is Added the Trayal of the Seven Bishops, with a Preface Shewing the Present Danger of Our Religion and Liberties ...*, published in 1713, began with the words 'Nothing teaches us more effectually to prevent future mischiefs, than the recital of our escapes from past dangers'.[4]

Recollections of the seven bishops were also cherished in the popular memory. In 1709, Archbishop Sancroft's ghost was said to have appeared. A popular account of the apparition described him as a bishop of 'unblemish'd character ... whose illustrious memory will ever be dear to all true Protestants, for the great honour he did the Church'.[5] It was another 12 years before the death of the longest surviving of the seven bishops, Jonathan Trelawny, who had been only 38 years of age during the trial. His funeral procession travelled from Chelsea to Cornwall in great pomp; the hearse was preceded by banners, his mitre and crozier. It was the passing of the last episcopal hero of 1688, and crowds lined the path of the procession.[6]

The memory of the heroic seven bishops continued to be commemorated during the eighteenth century. In 1727, *A Collection of Epigrams* included the well-known verse:

> True Englishmen drink a good health to the mitre
> Let our church ever flourish, tho' her enemies spite her
> May their cunning and forces no longer prevail
> And their malice, as well as their arguments fail
> Then remember the Seven, which supported our cause
> As stout as our martyrs, and just as our laws.[7]

It was a poem popular enough to be reprinted in the 1736 collection *The Flowers of Parnassus, or the lady's miscellany*. For a long time, the bravery of the seven bishops was used as an example to others; one clergyman recalled 'their memory is consecrated in every Christian English breast and is calculated to call forth, in all orders, some conspicuous instances of a pious constancy not less meritorious than theirs, in the same sacred cause'.[8] When Parson Hawker of Morwenstow sought an image with which to clothe the Cornish unrest of the mid-nineteenth century, he did so by alluding to Bishop Jonathan Trelawny's imprisonment, with the pledge that

> And shall Trelawny die?
> There's twenty thousand Cornishmen
> Shall know the reason why.

A hero of 1688 was called to mind to champion the cause of a later generation. Accounts of the trial of the seven bishops were printed regularly during the eighteenth century and sold well, especially the editions issued in 1735 and 1739.[9]

In April 1848, the *Eclectic Review* included a study of 'James II and the Protestant Bishops' which described the trial:

> Seven innocent men were arraigned at the bar, whose only alleged offence was the exercise of that right of free petition which had been esteemed the inalienable privilege of every Briton.[10]

The article pointed out that the weight of folklore (the commemoration of the Gunpowder Plot), public scandal (the Popish Plot) and law (the Test Act) pressed down on the people that 'recollections [of "Bloody Mary's reign"] made popery abhorrent to the people'.[11]

As recently as June 2005, on the anniversary of the acquittal of the bishops, Lord Tyler recalled his ancestor, Bishop Trelawny, and the way in which the news of the acquittal was received in Cornwall – with peals of bells, the discharge of town canons and consumption of great quantities of beer and wine. It was, claimed Tyler, an event which 'laid the foundations for the Bill of Rights on which our constitution is so firmly anchored'.[12]

Macaulay's heroes and villains

Like Tyler, Lord Macaulay regarded the Glorious Revolution as a turning point in English History. The 'Whig Interpretation of History' was born in the moment that Protestant progressive forces forged a constitutional monarchy from the ruins of James II's regressive Catholic policies. James, for Macaulay, represented unconstitutional despotism – a revival of, and throwback to, his father's divine right doctrine. In contrast, the seven bishops' protest at his use of the dispensing power and their acquittal in June 1688 was a triumph for the rule of law and the forces of moderation and progress. For Macaulay, the episode of the bishops' petition, imprisonment and acquittal encapsulated all that was bad about James and all that was good about the Glorious Revolution. Thereafter the royal prerogative was constrained by the rule of law and parliamentary authority. Macaulay was rarely a champion of the Church and its prelates, but faced with a villain as dark and reactionary as James, the seven bishops could not fail to be Whig heroes.[13] As this book will show, Macaulay's principal error was to mistake the conservative bishops for progressive and reforming constitutionalists.

Following Macaulay, two nineteenth-century authors considered the trial of the seven bishops. Agnes Strickland's *The Lives of the Seven Bishops Committed to the Tower in 1688*, published in 1866, showed how strongly supportive the seven bishops were of Charles II's policies and of James's succession.[14] Each of the bishops was a royalist, committed to the defence of the Church against Dissent and to the Tory anti-exclusionist agenda as the best means to preserve the Anglican monopoly. For these reasons they had advocated James's succession and welcomed it with high hopes. In some respects therefore Strickland endorsed Macaulay's view by indicating how far James had pushed away his natural supporters. But she also recognised that the bishops were essentially conservative in outlook. Strickland also showed that Sancroft and some of the others were by no means determined opponents of James, even after their release from prison. That five of the seven bishops subsequently became non-jurors and refused the oaths to William and Mary showed that they had not intended to challenge James's title to the throne or establish the 'progressive' constitutional settlement of 1660. In particular, Sancroft and Turner sought to keep James on the throne rather than accept either his abdication or a regency in 1689. Strickland placed much less emphasis on the minority opinion of William Lloyd and Jonathan Trelawny, who were so strongly offended by James that they became convinced supporters of his overthrow. Both were in contact with William during 1688, and probably shared the views of Bishop Henry Compton, who signed the invitation to William to come to England. Lloyd and Trelawny both quickly accepted the abdication of James. Thus the seven bishops did not represent a homogenous body of opinion towards James.

The second nineteenth-century study of the seven bishops was Herbert Lucock's *The Bishops in the Tower*, published in 1886. Lucock viewed the trial of the bishops through the wider lens of the reigns of Charles II and James II and the context of the struggle between Dissent and the Church. Like Macaulay and Strickland, Lucock criticised James and made the seven bishops the heroes of the Church and the Revolution. Sancroft and Ken were depicted as saintly and pious rather than astute and pragmatic in their motives, but Lucock suggested that they captured the popular mood. For Lucock

> the Bishops were foremost to see the danger and to realise the importance of making a resolute stand against his designs, and the nation gave them its full and cordial support...The imprisonment of the Bishops was an act in which the absolutism and Romeward policy of the restored Stuarts culminated; it was an event too which contributed immediately to their downfall.[15]

Lucock therefore discredited James, but, presaging later revisionism, saw his reign as part of a wider Restoration Stuart agenda that was shared with Charles II. Charles was as implicated in the use of dispensing power, and in the covert pro-Catholic policy. The only difference for Lucock was that, paradoxically, James was prepared to conciliate Dissenters while Charles, for the most part, permitted the persecution of Dissent.

More recent historiography has neglected the role of the seven bishops in the wider context of the Glorious Revolution. In contrast to Macaulay, Christopher Hill and Marxist historians of the 1960s saw the Glorious Revolution as 'the ultimate solidarity of the propertied classes', which averted 'the recollection of what had happened forty-five years earlier, when the unity of the propertied classes had been broken'.[16] Thus the aristocracy, gentry and mercantile classes ejected James because he threatened their interests, and William and Mary were installed because they would defend landed and commercial interests. Even recent accounts of the Revolution have offered a class-based analysis, identifying a 'lower class royalism' in England in 1685, a 'middling sort' resistance to James and an attempt by James to replace traditional Tory Anglican allies with a new class alliance based on the urban Dissenters' economic interests.[17]

In contrast, non-Marxist historians saw religion rather than landed and commercial interests as the centrepiece of the Glorious Revolution. Gerald Straka argued that during the summer of 1688 the most popular men in England were the seven bishops. After their acquittal they were the heroes of the populace; this is an important theme to which we will return. Straka also argued that the trial of the bishops became in effect a judgement on the dispensing power. Although the bishops were charged with sedition, they could not have been charged with it unless they had openly denied the dispensing power. It was the central weakness of James's position that he charged the bishops not for their refusal to obey him but on their challenge to his power to dispense with laws.[18] Straka also considered the degree to which the Glorious Revolution and the trial of the bishops reflected a united reaction to James and his policies. For the nation to see a united Protestant front of Church and Dissent against the King seemed to confirm Gilbert Burnet's tracts – issued from Holland – that the Church and Dissent had a common cause.[19] Straka also questioned the Whig interpretation of the Glorious Revolution as a reflection of unity among the English people and their leaders.

James and revisionism

Jonathan Clark, in *English Society 1688–1832*, sought further to correct the Marxist economic reductionist view.[20] Clark argued that the Revolution of 1688 secured the Anglican hegemony and aristocratic dominance against a Catholic monarchical bureaucracy. In effect, 1688 preserved what 1660 was supposed to have established. In this respect, while Clark rejected the Marxist view, he endorsed its implications; for both Clark and B. W. Hill, the Glorious Revolution was a conservative continuity rather than radical and revolutionary.[21] It preserved financial and religious interests. Kathleen Wilson claimed that the 'new orthodoxy' of 1688 is that 'the series of events once heralded as the foundation of modern parliamentary democracy is now presented as but a troubled and confusing hiatus in patrician politics, unrelentingly "conservationist" in ideological and political effect, in which Whig and Tory leaders managed to rid themselves of an unacceptable monarch without recourse to the political or ideological extremism of Charles I's reign'.[22] Significantly Jonathan Clark – in the second edition of his *English Society 1660–1832* – chose a portrait of the seven bishops for the frontispiece. He argued that the effect of James's Declaration of 1688 was to unite those who had been divided. Low Churchmen like William Lloyd were drawn to Sancroft and the other bishops. Clark also showed that the strength of opinions about passive obedience and resistance to rulers, after the accession of William and Mary, was unable to hold the alliance of High and Low Church together. Thus the alliance of forces against James was a fleeting one in the summer and autumn of 1688.[23]

Historians have also sought more comprehensive meta-explanations for the Revolution. Foremost among these is the role and responsibility of James for the events of 1688. J. R. Jones, in particular, challenged Macaulay's bleak view of James, adopting a revisionist account. Like Lucock, Jones felt that James's policies were, in many ways, no different from his brother's; where James differed from Charles was in his personal qualities and ineffectiveness in the handling of crises. Nevertheless, for Jones, James was a realist whose policies could have succeeded. His instruments of 'repression' were established Tudor and Stuart governmental systems and procedures, such as the Ecclesiastical Commission. Jones also paid considerable attention to James's attempt to pack Parliament in 1687–8, which he saw as a continuation of Charles II's policies and intimately connected to the Revolution in the provinces; whereas previously historians had seen the Revolution as principally a metropolitan event. This emphasis on a national Revolution gave it an important social dimension which historians had not previously explored. Jones also suggested that James had allied himself with the new urban middle classes against the aristocratic and landed interests; in this, James was progressive and innovatory whereas his opponents were forces of conservatism. Moreover Jones argued that, far from seeking to impose a Catholic tyranny, James was genuinely hoping to establish religious emancipation for Catholics.[24]

Jones also claimed that 'James's policies were realistic in the context of the general development of extensions of systems of absolutist power in government, principally in France, but also in Sweden and the German states'.[25] More recently, John Stoye has argued that James's prosecution of the seven bishops coincided with other European contests between Church and sovereigns. In France, Louis XIV was challenging the pope's right to interfere in French diocesan appointments, and the election of the elector-archbishop of Cologne was contested between the German states.[26] The way in which the Revolution of 1688 has been viewed as a European event, largely initiated and led by William, has been questioned by those historians who view it as an exclusively domestic British event.[27] Nevertheless, John Miller, J. R. Western and Barry Coward have asserted that James's fate was principally determined in Europe, and without William of Orange there could have been no Revolution in 1688.[28] Lucile Pinkham also emphasised the centrality of William as prime mover of the Revolution, arguing that the Revolution was not an impulsive act by William but the culmination of long-harboured plans. Thus the 'invitation' of the summer of 1688 was the pretext rather than the cause of the invasion. But William, claimed Pinkham, wanted the crown for its own sake as well as to bolster his European ambitions.[29] More recently both Jonathan Israel and Jeremy Black have viewed the Revolution as largely a European event which could not have happened without William's intervention.[30] Tony Claydon has also emphasised that even the Tories in 1688 supported foreign intervention, showing a European dimension to Tory thought.[31]

There are some problems with revisionist interpretations of the Glorious Revolution. First, Jones seemed content to take James's conversion to religious toleration and concession to Dissenters in 1688 at face value, yet he conceded that, in Scotland and America, James showed he was far less tolerant in outlook. How much James's emphasis on toleration was a cynical means to an end is not considered in detail by revisionists. In fact, of course if James *was* – as Jones argued – part of a wider European monarchical movement, historians need look no further than Louis XIV's revocation of the Edict of Nantes for a model of James's larger objective. Jones's emphasis on James as an innovator and an ally of the urban middle classes is also questionable – certainly as far as James's sincerity is concerned. Until 1687–8, James was closely allied to the representatives of the Tory landed interest, in opposition to the Whig exclusionists. The refusal of the aristocracy to cooperate with his Catholic policies is what led James to pursue the idea of wooing the Dissenters among the urban middle classes (if we can identify such a group) through the reconstruction of urban borough electorates. There is no doubt that key elements in the opposition to James were aristocratic and conservative, which, like the seven bishops, sought to prevent change in religious policy. These forces challenged the monarchical claims to dispensing authority and aimed to conserve political authority in the Anglican and landed classes. However the problem with this as an exclusive explanation for the Glorious Revolution is that it ignores many of the conservative elements which also supported James, the High Church Anglicans being an obvious example.[32]

G. V. Bennett, like Lucock, argued that the Glorious Revolution was the culmination of a policy dating back to the dissolution of the Oxford Parliament in 1681. Thereafter, the quickening pace of the Tory reaction to the Exclusion crisis saw Whigs cast out of boroughs and Tories placed in charge of the magistracy. This move was strongly supported by Sancroft and the majority of the clergy. Moreover Sancroft and his Tory allies used the Ecclesiastical Commission of 1681 to advance clergy who shared their High Church Tory principles. Many clergy advanced by the

Commission were also close to James, including a number of the seven bishops. Like Jones, Bennett believed the collapse in the Tory alliance of Church and monarchy did not stem from James's aggressive imposition of Catholicism – indeed he claimed that 'there is no real evidence that James ever planned to impose Catholicism by force' – but from James's desire to repeal the recusancy laws and the Test Acts of 1673 and 1678. In this, Bennett felt James went too far, tipping the Tory Sancroft-supporting bishops into preaching against Catholicism. By 1686, James was so offended by Turner doing so that he could not bring himself to speak to Turner at a levee. Faced with stiffening resistance from the bishops, James eschewed the kind of 'charm offensive' his brother might have attempted, preferring to intimidate them into backing down, confident that the Anglican principle of passive obedience would oblige them to do so.

What Bennett chronicled was a sequence of brinkmanship and counter-brinkmanship from which neither the bishops nor the King could withdraw. James underestimated the political ability of the bishops. Saintly and pious Sancroft and Ken might have been, but they were also effective political operators; and Trelawny, Turner and White were well-connected, the latter two particularly with the London clergy, and were able to persuade them to sign up to opposition to the King. Bennett correctly argued that, throughout the brinkmanship, 'James was outmanoeuvred by the bishops at every stage', with James maladroitly mishandling the public's perception of his role and under-estimating popular affection for the Church, even among Dissenters. Bennett was also right in claiming that Sancroft and the others were inept in their assumption that they could manage the crisis and keep it within Tory bounds. After the bishops' acquittal, the politicians, who had been relatively quiet during the spring and summer of 1688, saw their chance, Sancroft lost the initiative and events ran away from him, to a conclusion he had neither foreseen nor sought.[33]

William Spellman also emphasised the nature of the trial as a publicity disaster for James fuelling popular opposition to the King. But he emphasised that the trial led to the 'rapprochement' between the Church leaders and moderate Dissenters as several joint conferences were held in July in order to discuss future cooperation. Sancroft even contemplated a comprehension act and issued an exhortation to clergy to 'have a very tender regard to our brethren, the Protestant Dissenters' – this was a sharp change of policy for Sancroft.[34] Roger Jones also took the view that 'more than any other single event, the bishops' trial convinced the English that William alone could deliver them'.[35]

James's rehabilitation

A significant collection of essays, published in 1989, considered whether the Glorious Revolution happened 'by force or by default?' In it, J. P. Kenyon and John Miller pursued Jones's revisionist view of James. Keynon claimed that, by twentieth-century standards, James's policy of toleration was 'right'.[36] This is a curiously anachronistic assertion, and it is perhaps the only part of James's policy that bears this latter-day construction. Certainly his packing of Parliament, his policy towards the universities and his use of the dispensing power would attract starkly contrasting judgements. Miller, in examining Halifax's *Letter to a Dissenter* of 1687, challenged Halifax's view that James's conversion to a policy of toleration was insincere. Halifax wrote that Rome 'doth dislike the allowing of liberty' of religion. Miller suggested that Halifax's view of Catholicism was that it was 'authoritarian, morally dishonest and cruel'.[37] The problem with Miller's challenge to Halifax's view is that there is little evidence to clear James of a cynical and dishonest commitment to toleration.[38]

Miller's important study, *James II, A Study in Kingship*, argued that James's principal problems arose from Sunderland's self-interested encouragement of royal ambitions and plans as a way of bolstering his position against the more extreme Catholic voices. James's second problem was his unshakable belief in the divine right of kings and of the monarchy as a sacred institution. James had seen his brother treated with disdain by Parliament and dispense with Parliament as a source of revenue, and Miller implied that James felt he could similarly dispense with Parliament in law-making. If his duty was to rule, his subjects' to obey. Moreover James had a strong sense of providentialism: his conversion, the Restoration, his succession (against the best efforts of the exclusionists) and the defeat of Monmouth seemed to confirm that God guided him. Miller also suggested that James did not believe in establishing religion by force, and that this would have been impractical anyway as he had a Protestant army and no heir. Of course by 1688 all this had changed: James had established a standing army, he had repeatedly intruded Catholic officers and sought to legitimise their commissions, and by 1688 he was anticipating the birth of an heir – indeed these were exactly the fears that James's opponents expressed. Moreover Barrillon, the French ambassador, thought that James believed the Church of England was so close to that of Rome that conversion of Anglicans – as he himself knew – would be easy.[39]

Miller also argued that it was unreasonable for Englishmen to equate Catholicism with absolutism and arbitrary government. What Miller overlooked is that Englishmen had had plenty of evidence of the equation of Catholicism with absolutist and arbitrary government. If they had not sufficient evidence from the events in Europe such as the revocation of the Edict of Nantes, the reign of James provided evidence enough. Louis XIV, much regarded as a model by James, made clear that his subjects all had to share his religion. Miller believed that James's expression of admiration of Louis should not be taken at face value, claiming 'I suspect James genuinely disapproved of the violent methods Louis used to convert the Huguenots'. But even Miller conceded that James suppressed the publication of books in which the persecution of the Huguenots was depicted. Miller agreed that 'it is not easy to see a consistent pattern in James's behaviour, but I think it is possible'. Miller claimed that James disapproved of Louis's persecution of Protestants, yet a few lines later Miller conceded that 'James came to regard Louis's friendship as vital to the success of his Catholicising measures at home'.[40]

Miller considered that James's softening attitude to Protestant Dissenters was heartfelt. Yet James only turned to the Dissenters when he felt betrayed by his Tory Anglican friends and the bishops. This was no coincidence. Having failed to obtain repeal of the Test Act and a strong Catholic army from Parliament and, failing to prevent the clergy from preaching anti-Catholic sermons, he turned to the Ecclesiastical Commission, the punishment of the Church and conciliation of Dissent. Moreover James's much-quoted dislike of persecution really applied only to persecution of Catholics. In the same way, Miller accepted that 'James's sudden commitment to progressive economic reforms might at first sight seem merely a propaganda exercise'. It seemed to be insincere because it was: James knew in 1687–8 that he needed the Dissenters' electoral support, he knew the Dissenters had commercial interests and therefore he granted them economic concessions.

However Miller and Jones disagreed on whether James's desperate last-minute attempt to call a Parliament in autumn 1688 was sincere. Jones believed they were; Miller thought James was grasping at expedient straws. Miller's pleas for James took him to extraordinary contortions. He claimed

> If James stretched his powers beyond constitutional limits, he did so because he could not achieve his objectives without doing so. In accusing him of trying to establish absolutism, his contemporaries

and later historians confused means with ends, treating James's abuses of power and central rather than incidental features of his rule.

In fact the people who did *not* confuse means with ends were the bishops. They saw that James was abusing the dispensing power, and indeed this was the focus of their petition. They saw his behaviour was at odds with his coronation oath. They also knew exactly the objective of James's absolutist religious tendencies. Miller's judgement on James's reign was that it was 'not one of Popish despotism foiled by William's providential intervention but of extreme political incompetence and sheer bad luck'.[41] But despotism and incompetence are not mutually exclusive; James was a despot, incompetent and unlucky. Nevertheless Miller has argued that 'the extent to which James had alienated his subjects was shown by the petition and trial of the seven bishops... somewhat to their surprise they found that they were popular heroes... [and that] the rejoicing at the bishops' acquittal strained the government's ability to maintain order'.[42]

It has increasingly become clear that the lay politicians during James's reign equivocated in a way that churchmen did not. Halifax, Nottingham, Clarendon and the others stood on the sidelines, embarrassed by indecision. Shrewsbury, for example, when he travelled to Holland, mortgaged his lands for 30,000 pounds. If the Revolution came off he could redeem the mortgage, if it didn't he would have the cash.[43] Others hedged their bets shamelessly: Churchill had begun his correspondence with William in July 1687.[44] In considering the position of other laymen, Alfred Havighurst claimed that

> with issues changing so rapidly, it is idle to expect consistency. Reresby's memoirs are a testimony to this statement. So is the career of Sir Robert Sawyer – co-drafter of the Exclusion Bill, then Attorney-General and prosecutor of the Whigs, and finally counsel for the seven bishops and a member of the Convention Parliament.[45]

The same can also be said of the seven bishops themselves, who found themselves resisting a king despite their commitment to non-resistance.

The new Whig Revolution

Robert Beddard, the doyen of Revolution historians over more than 30 years, has made an important contribution to our understanding of the Glorious Revolution and the trial of the seven bishops. Beddard emphasised that francophobia and economic decline in the 1670s and 1680s were the backdrop against which the Revolution was set. Beddard held that the Tory role in the Revolution has unduly attracted the attention of historians. He has argued that the importance of the Tories in inviting William to England, their passivity in the face of the invasion, the salvaging of Tory Anglicanism in the revival of providential divine right theory and the justification of the Revolution as a deliverance of Protestantism from Popery have all been overstated. Beddard claimed that the Whig role in the Revolution was more important. It was Whig contractualists who pushed through the abdication of James, and the Convention Parliament was dominated by Whigs who largely framed the Revolutionary settlement of 1689. In this respect the Revolution was a Whig achievement. Beddard has also pointed to the internal conflicts and contradictions in Toryism: Sancroft, for example, veered between passivity and activism, and after the acquittal of the bishops the majority of the Tories lacked the ability and leadership to regain the initiative. As Beddard claimed

> That he [William] had obtained the throne with the unqualified backing of the Whigs, even of 'some hot-headed persons' whom he could hardly restrain, explains how from the point of view of the monarchy the Revolution came to be, in Macaulay's telling phrase, 'a preserving revolution'.[46]

The debate on the Glorious Revolution has moved on to more general questions. Was the Revolution truly a popular rebellion? If it was not, if it was simply a bloodless coup d'etat in which there was no real use of force, can James truly be said to have been deposed? Daniel Defoe wrote in 1702: 'The Church of England took up arms against their King in [16]88, and did not cut off his head because they had him not.'[47] It is a central argument in this book that the bishops' trial made the Revolution a popular one, in which every parishioner in the country had a stake.

A central issue in these competing Whig and Tory critiques of the Glorious Revolution is the role of the gentry and aristocracy, the landed Tory interest, compared with that of the Whig-Dissenting and urban interests. The Tory view of the Revolution as an aristocratic coup d'etat, in which the nobility played the principal part, has been considered in detail by J. P. Kenyon. There is much evidence for this view. In 1680, there were just 151 peers and 'from 1685 onwards he [James] also conducted an all-out assault on the privileges and influence of the nobility, and the fact that this was largely incidental to his main purpose did not make it any the less resented'.[48] That James was prepared to sack his own brothers-in-law, Clarendon and Rochester, in January 1688 showed that he would marginalise not only the nobility, but even his own family for his religious policies. Between August 1687 and March 1688, James dismissed 18 lords lieutenant from 21 counties. Yet the role of the majority of the aristocracy was not central to the initiation of the Revolution. As late as July 1688, leading noblemen, such as Halifax and Nottingham, firmly believed that James's Catholic policies would fail of their own accord and equivocated in their support of 'regime change'. Moreover before the invasion, William of Orange received remarkably little support from the aristocracy, the 'immortal seven' who invited him to come were largely dispossessed nobles. Admittedly, once the Prince appeared in the West, even James's most natural supporters quickly went over to him, but Kenyon suggests that only 10 per cent of the nobility were active in the Revolution. But this, in some respects, misses the point. Men, like the Duke of Beaufort, who had ridden to defend the city of Bristol from Monmouth in 1685, were not willing to take to their horses in 1688 for James. Beaufort reported that the 'best men' were lukewarm toward James even as William invaded. It was this that broke James's nerve.[49]

In contrast, the trial of the seven bishops mobilised popular opinion and did so against James in a direct way.[50] John Carswell asserts of the popular support for the seven bishops:

> it is often said that the Revolution of 1688 was not a popular movement, yet it is questionable whether it could have dislodged one source of security and installed a new one without this rising of popular fever as one dramatic event followed another against the background of probably war. Religious zeal and suspicion of foreigners combined to produce a horror of Catholicism.[51]

It is one of the central themes of this book that the trial of the seven bishops mobilised public opinion against James in a way that no other event did.[52]

Tory views atomised

Mark Goldie has also added important elements to the debate on the Glorious Revolution, endorsing Beddard's view of it as a Whig event.

From the start of James's reign, Goldie argued, the clergy were in the forefront of resistance, and therefore the illusion was created of a Tory Revolution. But however much, in the reign of Queen Anne, there was a tendency to see 1688 as a Tory Revolution, it was 'an irreducibly Whig event'.[53] Similarly, while it is easy to see in the trial of the seven bishops, what Goldie called 'the climacteric of the Anglican revolution', as ushering in the reign of William III, the seven bishops were, for the most part, horrified by the Dutch invasion and the displacement of James. The Tory bishops wanted James back on their terms: a Tory Anglican regime, which they briefly achieved in October 1688. If James had not made his cynical, desperate concessions so late they might have succeeded. But the dissolution of the Ecclesiastical Commission, the restoration of the fellows of Magdalen College, Oxford, and the rest came too late either for it to have any effect or for James to be trusted. It was the Whigs and Dissenters who grabbed the initiative, while the Tories only considered the terms on which they might let James keep the crown.

Goldie also considered in detail the internal inconsistencies of Sancroft and the Tory Anglican position. In contrast to the usual assumptions, Goldie argued that Restoration Anglicanism permitted resistance to the King. The Anglican doctrine of passive obedience did not apply in the case of threats to the Church. Indeed James and historians have misunderstood the nature of passive obedience. Resistance to a Catholic monarch was clearly permitted by the Anglican principle of passive obedience, claimed Goldie, on the grounds of conscience. Sancroft and the seven bishops asserted their right not to insist on the reading of James's Declaration of Indulgence on the grounds of their conscience. For Anglicans, such claims of conscience were appeals to God's law; the issue was not the temporal legality of the Declaration for Liberty of Conscience, it was its divine legality. The Tory Anglican emphasis on conscience was on an Anglican public conscience. In origin, Sancroft's views were strongly anti-Erastian, they denied the supremacy of the state over the Church and posited an establishment made up of two institutions, Church and state, independent but mutually supportive. As Goldie argued, 'the trial in 1688 by an apostate prince of seven bishops of the Christian church was an occasion that demanded a fulsome enunciation of these principles'. Sancroft's draft speech (written for any occasion on which he might have been required to speak during the trial, but not used) showed that he sought to defend the Church against an absolute liberty of conscience and against repeal of the Test Act. When, after his acquittal, Sancroft opened discussions with Dissenters, Goldie suggested he was insincere, his conversion to toleration being 'only skin deep'.[54] In this respect, Sancroft and James shared a Tory Cavalier view of things.

Goldie also claimed that the imprisonment and trial of the seven bishops was a drama that highlighted the theme of martyrdom. Indeed he argued 'one of the most striking aspect of the Anglican resistance was its noisy self-dramatization as martyrs under persecution'. Trelawny and all the bishops revelled in their image as martyrs suffering for the Church. It is a theme of this book that the bishops managed their public image remarkably well. In prints and medals they were depicted as saintly men facing the same persecution as the Protestant martyrs under the last Catholic ruler, 'Bloody Mary'. Prints of the seven bishops were distributed far and wide, 'never have the images of English clerics been so widely dispersed.' William Sherlock in November 1688 spoke of the bishops as facing 'the jail and the stake'.[55]

This well-orchestrated campaign led to a feature of the Glorious Revolution that is controversial, the involvement of the wider population. It is quite clear that the vast majority of English men and women experienced James's policies and the opposition to his regime through his religious acts. They were relatively untouched by changes to borough charters and franchises, or by the ejection of lord lieutenant or the creation of a standing army. The way in which the Glorious Revolution became a national revolution was through the trial of the bishops. Sancroft, a firm adherent to the cult of King Charles I as martyr, welcomed the tribulations of 1688 as a contemporary echo of that martyrdom. In a conclusion redolent of J. R. Jones, Goldie asserted that 'by the perversity of their witness on behalf of the exclusive claims of their own Church, the Anglicans, like the early Christians, vexed and antagonised an empire otherwise generally predisposed to tolerance'.[56] But such a judgement only holds if James was truly committed to toleration, as Yvonne Sherwood has asserted, 'it would be naïve in the extreme to see James II/VII as a campaigner for religious toleration'.[57] Either way, the bishops knew the power that the rhetoric of martyrdom would have with the Dissenters. It did not take a latter-day Foxe to show to the Dissenters that by resisting James and being tried by him created a community of interests with Anglicans.

Continuity and change

W. A. Speck claimed that the revisionism of James's rule has gone too far, a view shared by Tim Harris.[58] Harris also challenged Jones, Israel and Black when he claimed 'external factors alone cannot explain why James's regime toppled...it was domestic political turmoil that in the main was responsible for bringing James down'. Harris was more critical of James than most recent historians and was in no doubt that 'to all intents and purposes James did try to establish Catholic Absolutism in England'. He also argued that 'to claim that the Revolution came essentially from above and was the result of external factors is seriously misleading'. Like Goldie, Harris advanced the view that passive obedience has been misunderstood by historians and that it always contained a contractual element and therefore Tory Anglicans only understood their duty in terms of how to behave to a just and good King. Harris suggested that when the bishops went to James with their demands in October 1688 and he conceded all they sought: 'there was an Anglican Revolution in the autumn of 1688, in other words, which preceded the ultimate Williamite Revolution'. The paradox of the bishops' position is clear in Harris's account: 'James ended up being resisted by people who believed in non-resistance, and held to have done wrong by people who believed the King could do no wrong.' Ultimately, for Harris, the Revolution was conservative, even though he regarded the Revolution as one 'from below' and even though it was a more radical change than happened in 1641–2.[59] In contrast, Steven Pincus has claimed the Glorious Revolution was 'the first modern revolution' and was consciously modernising; both James and William sought progressive changes. Pincus claims that the Revolution effected radical change in the English state, economy and Church.[60]

Michael Mullett's attempt to synthesise work on James II and the Revolution places considerable weight on James's personal flaws – as does Tony Claydon who called James 'a pig-headed bigot'.[61] Mullett also recognised that James's deposition owed much to the trial of the seven bishops. Whereas Halifax's *Letter to a Dissenter* may have worried James, he was much more alarmed by the popular reaction to the bishops' acquittal. But Mullett concluded with a paradox. While he recognised that 'the seven bishops showed the King could be opposed without being dethroned' and therefore William played an indispensable part in the Revolution, he also asserted that the lesson of 1688 was that 'no seventeenth century European monarchy was stronger than the landed social and ecclesiastical elite'.[62]

The seven bishops

Against the backdrop of such historiography, this study of James II and the trial of the seven bishops provides an opportunity to consider their importance in the Revolution of 1688 afresh. The Revolution was undoubtedly an Anglican revolution first and foremost. People chose their Church over their King. One of the key features of the contest between the King and the bishops was the widespread use of propaganda by James and the bishops in an overt attempt to win over popular support. James's propaganda was pushing against the prevailing anti-Catholic tide of Anglican and English traditions and lore and lacked both the credibility and the finesse of that of the bishops and their supporters. The bid for popular support shows how important public opinion was and how much the events of 1688 featured in the public mind. This may not have been a popular revolution in which the mob surrounded the palace, but, for most Britons, the palace was occupied by the enemy. That conclusion could not have reached without the trial of the seven bishops. Public opinion – at least in London – was sympathetic to Henry Compton when he was suspended from the diocese of London, and it was astonished by James's treatment of the fellows of Magdalen College, Oxford. But these events did not bring people to the point of the treasonous passivity which they reached by the autumn of 1688. Nor did the ejection of JPs and lords lieutenant, or the *quo warranto* manipulation of borough representation, bring people to the sticking point. What ensured that James was spurned as a tyrant and William welcomed as a saviour was the imprisonment and trial of the seven bishops. As diocesan bishops, they brought a territorial actuality to their persecution; the people of Canterbury, Ely, Chichester, Bristol, St Asaph, Bath and Wells and Peterborough found their religious leaders gaoled and on trial. The people of these and other dioceses saw they had a King who had broken his coronation oath, who had declared war on his own bishops and abandoned the rule of law. Thus the popular Whig revolution of 1688 could only have happened after the trial.

The same is true of the Tory coup d'etat. From February 1685 to July 1688, politicians and aristocrats who viewed James with ambivalence had shown little spinal fortitude. Of Danby, Halifax, Nottingham and Clarendon, the best that can be said is that they were cautious. But caution changed nothing, and conceded everything to James. If the trial showed anything it was that incautious bishops, bold judges and a courageous jury could change the political landscape – not least the legitimacy of the dispensing power. Without the echoes of the shouts of joy ringing in their ears would the 'immortal seven' have sent their invitation to William on the evening of the acquittal? The acquittal made them bold in defence of the Church and in defiance of the King.

The trial also provided evidence that James had broken his contract with his subjects. While Whig theorists later made a broad and generalised claim, in reality James had broken a very specific contract in the form of his coronation oath, to defend and protect the Church of England. When Harry Dickinson wrote that 'it was not easy to find evidence of resistance in 1688 or to prove that James II had been deposed for breaking the original contract', he overlooked James's breach of his undertaking to the Church and the bishops petition and refusal to read the Declaration.[63]

The trial of the seven bishops was James's El Alamein; before it, he had not been defeated; after it, he had no successes. It gave courage to the disparate groups that James had offended to join in opposition to him. Would the Glorious Revolution have happened without the trial? Counterfactual speculation is a perilous path. But it seems reasonable to suggest that without the display of popular support for the Church against the King, the politicians would have continued to waver. Without a popular slogan – coming to defend the liberties of the Church – William might have been viewed as a Dutch enemy, invading principally to advance his own interests. Without the trial and its propaganda war, the aristocracy would not perhaps have seen how irrelevant they had become to James – for whom Dissenters and boroughs could alone secure him a majority in the Commons. In short, while it is not the case that without the trial there would not have been a Revolution, it is certainly the case that the treatment of the seven bishops made the Revolution bloodless and easy; they made it glorious. Though the seven were by no means of one mind after the flight of James, they were conscious of the challenge they had made to him. It made them the midwives of the Glorious Revolution.

1
The Bishops: Unlikely Revolutionaries

The seven bishops were the most implausible revolutionaries; in many respects their backgrounds, education and churchmanship made them much more likely to support James than oppose him. One of the writers during the contest between the King and bishops commented that James 'had contributed so much to the placing of most of you in your episcopal Chairs'.[1] John Miller wrote of the events of 1687–8: 'the changes...happened in spite of, not as a result of, the prevailing constitutional theory. Men were forced by immediate political circumstances to take actions which directly contradicted their fundamental constitutional beliefs.'[2] This exactly sums up the position of the seven bishops. They were all natural supporters of James who found themselves acting against him. The seven were the builders, literally and metaphorically, of the Restoration Church; and they were men of principle.[3] They were also men of subtle but significant differences, and while history has treated them as a single group it is important to recognise that the seven bishops were not of a single mind. Moreover, the root of their resistance to James can be traced to elements in the bishops' early lives and careers.

Family and background

The bishops were by no means a homogenous group socially; they came from a range of social strata. The highest born were Trelawny and Turner. Jonathan Trelawny came from an ancient Cornish landed family and was born deep in the royalist cause: his father and grandfather had taken up arms in the cause of Charles I. By the end of the Civil War, the Trelawnys were saddled with debts and fines which left them seriously impoverished. Francis Turner was the son of Thomas Turner, a chaplain to Charles I, and Margaret, daughter of Sir Francis Windebank, Charles I's Secretary of State. Inevitably therefore the Turners were committed royalists who suffered the full privation of those who had been close to the King. The family went without food on occasion and often lacked somewhere to live. The 'middling sort' was represented among the bishops by William Sancroft, who came from a Suffolk family of yeomen farmers who could trace their ancestry back to the Middle Ages; Thomas Ken was the son of a Welsh lawyer and William Lloyd the son of a parson. Two of the bishops came from poorer homes: John Lake may have become a Tory cavalier, but he was born the son of a Halifax grocer and Thomas White's parents could not afford to send him to university so he went as a sizar, a servant who waited on other students.

Education and training

Naturally all the bishops were educated at the universities, Sancroft, Lake and White at Cambridge; Turner, Ken, Lloyd and Trelawny at Oxford. Most were accomplished scholars. Sancroft, whose talents meant he could have chosen any profession, opted for the Church. He wrote to his father, in September 1641:

> I have lately offered up to God the first-fruits of that calling which I intend, having common-placed twice in the chapel; and if, through your prayers and God's blessing on my endeavours, I may become an instrument in any measure fitted to bear His name before His people, it shall be my joy and the crown of my rejoicing in the Lord. I am persuaded that for this end I was sent into the world, and therefore if God lends me life and abilities, I shall be willing to spend myself and be spent upon the work.[4]

Graduating in 1637 and incepting as MA in 1641, Sancroft was elected a fellow of Emmanuel in 1642. His election troubled him; in April 1642, when he had been left a small income, he wrote to his father:

> my quaere is, whether this assignment (though but in trust), especially if the trust be not mentioned in the instrument, will not invest me with such an estate as will disable me from taking this preferment in the College. That nobody knows of it I weigh not, for I desire more a thousand times to approve myself to God and my own conscience, than to all the world beside. If it be not done, I pray, sir, think upon it before you do it; if it be done, and you find it will touch upon the

statute, let it be undone. I would not be too scrupulous nor too bold with my conscience. If it be a needless scruple, I had rather show myself to have no law than no conscience.[5]

Whether to follow the law or obey his conscience would also be Sancroft's dilemma in 1688. For others, the decision to enter the Church was a more pragmatic decision; for example, as a second son, Jonathan was not expected to succeed to the Trelawny estates and baronetcy and consequently sought ordination.

Education in the middle years of the seventeenth century was an intensely political activity; the universities were subject to regulation by Parliament, and fellows and students expected to commit themselves to parliamentary principles. All of the bishops who were at the universities in the 1640s and 1650s had to face the hard decision whether to submit to Parliament. Sancroft's election to a fellowship at Emmanuel coincided almost exactly with the outbreak of the Civil War. He was firmly on the side of the King, describing Parliament's actions as 'arbitrary'. But Cambridge and later Oxford were forced to accept the parliamentary Covenant. The Earl of Manchester visited Cambridge and forced Dr Holdsworth, the master of Emmanuel College and vice-chancellor, out of his offices, and imprisoned him for his defence of the King. Sancroft, who was a friend of Holdsworth and made no secret of his resistance to Parliament, was sufficiently junior to be overlooked in this first attack on the University. Sancroft wrote to Holdsworth:

> We live in an age in which to speak freely is dangerous; faces are scanned, and looks are construed, and gestures are put on the rack, and made to confess something which may undo the actor; and though the title be liberty, written in foot and half-foot letters upon the front, yet within there is nothing but perfect slavery.[6]

This was a useful preparation for the contest in 1688 when a similar fear of openly expressed views confronted the bishops and clergy, and Sancroft had to be careful to whom he spoke about his growing concerns at the actions of James II.

When it became clear that Sancroft could not delay the choice between signing, or refusing, the Parliamentary Association, he wrote with perhaps a conscious reference to Cranmer:

> And what then? Shall I lift up my hand? I will cut it off first. Shall I subscribe my name? I will forget it as soon. I can at least look

up through this mist, and see the hand of my God holding the scourge that lashes; and with this thought I am able to silence all the mutinies of boisterous passions, and to charm them into a perfect calm.[7]

Despite his lifelong interest in martyrdom, Sancroft did not seek out confrontation, preferring to remain unobtrusively in Cambridge. He refused to abrogate his conscience, and claimed to be motivated only by 'God's voice in my soule'.[8] A combination of tact, wiliness, absence from Cambridge during bouts of ill-health, and friendship with some of the key parliamentarians in Cambridge, such as the new Puritan master of Emmanuel, Anthony Tuckney, meant that Sancroft managed to evade both signing the Association and ejection from his fellowship.[9] Indeed in 1648 he proceeded to the degree of BD.

Sancroft was appalled by the execution of Charles I, who he claimed was 'the best Protestant in these kingdoms'.[10] In Sancroft's view it was a horrifying act of rebellion that went beyond the raising of arms by Parliament. This high, sacramental view of kingship was a feature of his thought which meant that when James II assumed the throne he did so with Sancroft's natural support and loyalty. It also focused Sancroft on an act of martyrdom and victimhood with which he was to become preoccupied. With the formation of the Commonwealth, Sancroft's ejection could not be delayed much longer. Finally in July 1651 he was expelled from his fellowship. Sancroft had been left a small estate to which he retired. It was a solitary life but one which Sancroft's advanced monastic sensibilities enjoyed.

John Lake graduated in 1642, just as the Civil War was breaking out, and he quickly showed that he was an unequivocal royalist. He was arrested, together with other ardent young Cambridge royalists, by the parliamentary commissioners, for refusing to take the Covenant. Lake was imprisoned but escaped to Oxford and joined the royalist forces; subsequently he served in the army for two years. He distinguished himself at the siege of Basing House by his intrepid conduct in several successful sallies. Lake was fortunate, however, in escaping the massacre of the garrison in revenge for its refusal to surrender and was left for dead at the siege. He was also active at Wallingford, one of the last towns to surrender after the capture of the King. When the royalists were defeated, Lake fled to Halifax where he obtained holy orders secretly from the Bishop of Ardfert. He was ejected from his living in Halifax for refusal to take the engagement to the Commonwealth and moved to Oldham where he quietly acted as curate in the parish of a royalist supporter. For a time he was able to hold out against the Manchester Presbyterian classis, but by 1654 he was also ejected from Oldham.[11] Some of the other bishops were less willing to make a commitment as strongly as Lake. Thomas White, despite having skills as a boxer,[12] a talent he was rumoured later to have employed against Bishop Cartwright, kept a low profile during the Commonwealth period, when his whereabouts were unclear.[13] In Commonwealth Oxford, Thomas Ken joined a quiet group of devout Anglicans, led by John Fell, for worship. In defiance of puritanical disapproval, Ken established a musical society, at which he occasionally sang and performed on the lute, viol and organ.[14]

Others were even more willing to accommodate the Puritan establishment in the Commonwealth. During the Civil War, William Lloyd seems to have shuttled between Berkshire and Oxford, avoiding the surrender of the town but obtaining his MA in 1646. In 1648, he was ordained deacon by Bishop Skinner of Oxford and priest by Bishop Brownrigg of Exeter in 1656. But he accommodated the Presbyterian regime, though he also visited the royalist court during its exile in France. He acted as a private tutor and also appears to have been involved in a fraud in which he duped Oxford Presbyterians into believing that a merchant disguised as a Greek Orthodox patriarch was the real thing. Francis Turner was educated at New College, Oxford, graduating in 1659. In order to do so, he had taken the engagement to the Commonwealth – being the only one of the seven bishops to do so.

Early preferment

The Restoration in 1660 brought about a dramatic change in fortunes for the bishops. In November 1657, William Sancroft travelled in Europe and visited the royalist court in exile. He travelled on to Italy and at Padua he briefly entered his name as a student at the university. Arriving in Rome, in May 1660, Sancroft received news of the Restoration, and returned to England early in the autumn. He arrived too late to benefit from the first allocation of livings to those who had been in Holland with the exiled court. Nevertheless, in November 1660, Sancroft was chosen to preach the consecration sermon in Westminster Abbey on the appointment of his friend, Dr Cosin, to the bishopric of Durham, and of six other bishops.[15] This was his first public appearance, and his sermon was admired as an expression of the apostolic character of the Church of England. The result of the sermon was that he was immediately chosen by Bishop Cosin as his domestic chaplain; and the next year, on the recommendation of Archbishop Sheldon, he was appointed as a royal chaplain by Charles II.

Sancroft was not present at the Savoy Conference, which unsuccessfully sought to broker a compromise agreement on liturgy and doctrine between Anglicans and Presbyterians. But, through Cosin's good offices, Sancroft assisted in the revision of the liturgy. Sancroft's interest in martyrdom again came to the fore when he was one of the three supervisors of the new Book of Common Prayer, which contained the liturgies for the Restoration and the martyrdom of King Charles I. This laborious task gained him the gratitude of the King, a DD from Cambridge, and, more significantly, the 'golden rectory' of Houghton-le-Spring in Durham and a canonry of the Cathedral.[16]

In August 1662, Sancroft was elected master of Emmanuel College, Cambridge, and proposed major reforms and rebuilding; but since he was only master for three years few of these were completed. However, Sheldon ascribed Emmanuel's conversion from a troubled college into a pillar of the Anglican establishment to Sancroft's leadership and guidance.[17] This was to begin a period in Sancroft's career in which he earned a reputation as a builder; physically he repaired the fabric of the buildings for which he was responsible, but he also rebuilt the Restoration Church as a pillar of the Restoration state. Sancroft also attracted the approval and patronage of leaders in the Church and state. It is clear that Sheldon saw in Sancroft a man who could carry forward the Tory Cavalier Anglican agenda.

Sancroft's reward came in January 1664 with his nomination to the deanery of York. He held the deanery for just nine months. Here too he went to great expense to repair the ruined deanery, and rebuilt some portions where repairs were impracticable. In November 1664, the King appointed him to the deanery of St Paul's on the recommendation of Archbishop Sheldon and Bishop Henchman of London. Sancroft devoted himself to the task of repairing and restoring old St Paul's Cathedral.

St Paul's had been badly damaged during the Commonwealth, and used as a barracks and a stable for the parliamentary cavalry. In 1666, the Great Fire of London, completed the work of destruction begun by the Commonwealth regime, and left old St Paul's a scorched and blackened shell. In October 1666, Sancroft preached a fast sermon before Charles calling for national repentance of the sin that had caused the plague and fire.[18]

Attempts to repair the devastated building were soon abandoned, and Sancroft presided over the commission to Wren to rebuild the whole structure, contributing £1,400 as well as devoting an annual sum from the chapter funds. It was an arduous time for Sancroft, as London was almost bereft of churches. Worse still, Paul's Cross, Charing Cross and the ancient cross in Chepe, which had been sites around which the poor congregated – deterred from attending public worship by the cost of pew rents – had also been demolished by the Commonwealth. Sancroft saw the urgent necessity of hastening the rebuilding of St Paul's, if only to provide the people of London with a place of worship. Sancroft also rebuilt the deanery of St Paul's at a cost of £2,500.[19] A vital part of his role was to solicit the proceeds of the coal tax to pay for rebuilding the parish churches of London, and he helped draft the legislation to unite those city parishes where the churches could not be rebuilt.

At the Restoration, the dean of York presented John Lake to the vicarage of Leeds. But he met with so much opposition from the Puritan party that Lake had to call in a company of soldiers to secure his induction into the church, the doors having been barred against him by some of the more violent of his congregation. Lake preached his first major sermon at York, which was so impressive that the dean sent a copy to Gilbert Sheldon, bishop of London. Sheldon was looking for able young clergy and appointed Lake to the rectory of St Botolph's, Bishopsgate, in May 1663. A year later he was also made prebendary of Holborn in St Paul's. During this time, Lake formed a friendship with Sancroft based on their work to rebuild churches after the Fire of London. Lake was also made prebendary of York, where he acted as magistrate. He showed his determination to enforce church discipline, ordering the abolition of the custom of permitting the congregation to walk about the aisles of York Minster and talk during services. This led to an outbreak of violence in October 1680, on Lake's installation as archdeacon of Cleveland.[20]

Despite his disappearance during the Commonwealth, in July 1660 Thomas White successfully petitioned the King for the vicarage of Newark, suggesting he had a reliable loyalist record. Six years later, he was preferred to the rectory of Allhallows, London, where he was a popular preacher and attracted a congregation running into thousands.[21] White came to Charles II's attention after an altercation in which he defended the bishop of Rochester during a scuffle with an ill-mannered trooper. In 1683, Charles appointed him archdeacon of Nottingham, this signalled White's movement into the royalist sphere; and he was made an Oxford DD, an honour conferred on only the most reliable Tories.

At the Restoration, despite Francis Turner's acceptance of the Commonwealth, he followed his family in a return to royal employment. His father was restored to his old preferments and Francis was appointed chaplain to Anne Hyde, the Duchess of York, James's first wife. In 1664, he became rector of Therfield, Hertfordshire and in 1666 was elected a fellow of St John's College, Cambridge. Three years later he was appointed to a prebend of St Paul's Cathedral, where he met Sancroft.[22] In 1670, Turner became master of St John's College, Cambridge and completed the rebuilding of the college. Like Sancroft and Lake, he was a keen builder, reconstructing his rectory at Therfield at a cost of £1,000.

About this time Turner, like White, gained a reputation as a preferment seeker. Andrew Marvel referred to him as a 'youth who treads always upon the hem of ecclesiastical preferment'.[23] But it was his Tory loyalism that gained Turner the advancement he sought. When, in the wake of the Popish Plot, the Duke of York was banished to Scotland, he chose Turner to go with him as his chaplain. Despite his Catholicism, James always maintained an Anglican chaplain because he knew that not to have one would antagonise the Church. Few clergy could be more sympathetic than the son of his father's chaplain, who had travelled with the King during all the troubles of his reign.

Turner endured the Duke's exile in Scotland, where James treated him with great kindness and even confidence; moreover James promised to use his influence with the King for Turner's appointment as dean of Windsor and royal almoner when the offices fell vacant. Turner was naturally impressed by the Duke and wrote to Sancroft that James was

> the best master... My lord, for want of other company I have more discourse with the Duke than otherwise should come to my share, and upon all occasions I find he places his hopes altogether upon the episcopal party, and mainly upon the bishops themselves, your grace especially; wishing and desiring that your grace will take all opportunity of encouraging the King (that was the Duke's own word) to be steady in well chosen resolutions, and laying before his Majesty how fatal a thing it would be now to trace back again the ground he has gained, and how mighty safe to stick by his old friends and the laws.[24]

While in Edinburgh, James broke the monopoly of the widow of the King's Printer on the printing of the Book of Common Prayer, which had prevented its widespread distribution. This was regarded as evidence of James's good intentions to the Church of England. James also supported the efforts of Mr Gordon, an Anglican missionary in the colony of New York. In effect, James sought to reassure the Tory party in the Church that he was protective towards Anglicanism. Turner, who also acted as James's channel of communication to the Scottish episcopal hierarchy, ensured the Duke was not embarrassed by demands for promises of toleration from Scottish Catholics and Presbyterians. In Scotland, Turner also reported that James placed all hopes of his succession on the bishops and the Church and he saw the Church as an ally against his Whig exclusionist enemies.[25]

Thomas Ken delayed his graduation until the Restoration and was subsequently ordained. He received the degree of MA early in 1663, and was appointed chaplain to Lady Maynard, who also presented him with the living of Little Easton, Essex. Ken's contacts with Lord Maynard and with George Morley, who had returned as Bishop of Winchester, meant he had access to patronage. In 1665, he was appointed chaplain by Bishop Morley of Winchester, resigning his living in Essex, and taking up residence with Morley. Two years later, Ken was appointed rector of Brightstone, on the Isle of Wight, before Morley advanced him to the rectory of East Woodhay and a prebend of Winchester.

At the Restoration, William Lloyd also received the first of a series of preferments, which continued to fall on him for 40 years. The first was a prebend of Ripon; then a prebend of Salisbury, and a resident canonry there followed. In 1668, he became vicar of St Mary's, Reading and dean of Bangor in 1672. He also held the archdeaconry of Merioneth, and was chaplain in ordinary to the King.[26]

Jonathan Trelawny was younger than the others and only entered Christ Church, Oxford, in 1668, by which time the Restoration had repaired much of the Trelawny patrimony. Whilst he was pursuing his scholastic career, his father was comptroller of the household of the Duke of York and a colonel in the Duke's regiment of horse. Anne Trelawny, Jonathan's sister, was a lady-in-waiting to Princess Mary before her marriage to the Prince of Orange, and went with her to Holland as maid of honour. Jonathan graduated in 1672 and was ordained four years later. In 1677, his connections earned him a royal chaplaincy and nomination to the Cornish livings of Calstock and St Ives. In 1681 – his older brother having died – he inherited his father's baronetcy but also his debts. So he had to seek after preferment. He also had civil responsibilities as Vice-Admiral for Cornwall and a magistrate. In consequence, Sancroft had doubts about Trelawny's suitability for ecclesiastical preferment.[27]

Consecrations

Their efforts in support of the Church and the Duke of York led to appointment to bishoprics for the seven. The most remarkable of these was Sancroft's. On the death of Gilbert Sheldon in November 1677, the King – who, during the building of St Paul's Cathedral, had become closely acquainted with Sancroft – decided to appoint him as the next archbishop, even though this would mean advancing him over the heads of all the other bishops. There is no doubt that Bishops Henry Compton of London and Nathaniel Crewe of Durham felt themselves both more senior and better qualified for the primacy. Lord Thomond joked about Sancroft's appointment with the Bishop of Durham: 'My lord, you have all been played a Newmarket trick; but you see, God Almighty's rule doth sometimes hold. He has exalted the humble and meek, and kept down the mighty from the seat.'[28] When Sancroft asked permission to decline the archbishopric, the King joked, 'You must take it, as you are quite homeless, for I have given away your deanery of St. Paul's over your head to Dr. Stillingfleet'.[29] Sancroft gave in, and his consecration as archbishop of Canterbury took place on 27 January 1678.[30]

The other bishops were not as diffident in their attitudes to mitres. Through the friendship of Lord Derby, who was sovereign of the Isle of Man, John Lake was nominated to the bishopric of Sodor and Man, and consecrated in December 1682. It was a very poor see, and Lake had to make a significant financial sacrifice, as he had to give up his much more lucrative preferment at York. Lake regretted his decision to accept the see, indeed he tried to escape it. In March 1684, before setting out for the Isle of Man, he wrote to Sancroft:

> The late Bishop of Carlisle being dead, my lord of York, ex mora motu, hath prompted me to put in for the bishopric, and hath promised all his favour and furtherance, which I took myself obliged to accept.[31]

Lake only remained on the island for a year because, in the summer of 1684, Sancroft used his friendship with the King to remind him that he had much enjoyed one of Lake's sermons. Accordingly, Charles granted Lake translation to the vacant diocese of Bristol. This was hardly the better preferment for which Lake had hoped, and Lake was not entirely overjoyed at the value of his new diocese. He wrote to a friend:

> I shall take the bishopric of Bristol, if it falleth to my share, not only contentedly, but joyfully and thankfully, if it can be but so contrived

that I may be entitled to another year's profits of the bishopric of Man, which I shall be perfectly at or before Michaelmas next, that is so soon as the harvest is cut down; but otherwise I shall be a great loser by the one, and I doubt a greater by the other.[32]

At Bristol, Lake sought to repair the cathedral and, in a move which reflected his High Churchmanship, supported the introduction of a weekly communion service. Sancroft used Lake to restore ecclesiastical order and discipline in the diocese of Lichfield and Coventry, and afterwards in that of Salisbury. In both cases Lake acted as Sancroft's commissionary in executing a metropolitan visitation. Consequently, Sancroft agreed to Lake's translation in 1685.

Thomas White was made domestic chaplain to Princess Anne on her marriage to Prince George of Denmark. As Princess Anne occupied an important place in the succession to the throne, after her father and childless elder sister the Princess of Orange, the appointment of White showed significant trust. Queen Anne's later commitment to the Church of England, and her High Church leanings, were seen as evidence of White's chaplaincy and Bishop Compton's tutorship. As Sancroft's Tory agenda for the defence of the Church advanced, he proposed White for the see of Peterborough in 1685. On Sancroft's behalf he conducted the metropolitical visitation of Lincoln in 1686. Like Sancroft and Lake, White was a stern disciplinarian, in his own diocese he was shocked to discover how many pluralities there were. After serious reflection on the practice, he submitted to Sancroft a list of cases which he regarded as serious abuses, and asked for his help in reforming them:

1. When one man has from one to three or four and five curacies to supply, and they do not altogether make up a competent livelihood? Many of which are I believe to be found in the northern parts of Lincolnshire.
2. Where one man holds a curacy and a vicarage or rectory, and perhaps lives at neither, but yet supplies them both by turns?
3. Where one man has two benefices with cure, and devolves them both upon curates to supply; he himself not being detained from them by any other employment, but chooseth some city or great town to reside in for his secular convenience?
4. Where pluralities belong to a residentiary in some cathedral church, and are supplied by curates, the incumbent never residing, or hardly ever seeing his parishes for several years.[33]

After his sojourn in Scotland with James, Francis Turner returned to Court, and took up residence in his prebendal house in London. In accordance with James's promise, he was appointed dean of Windsor and Almoner to Charles II early in 1683; and in July the same year, he was recommended to the King by Sancroft for the vacant see of Rochester. There were mutterings that Turner was too young for such preferment, but Sancroft was keen to have someone with Turner's political skills on the bench of bishops. On the death of Peter Gunning, Turner was translated to Ely, confirming both Sancroft's and James's favour towards him.

Thomas Ken was something of an ascetic, always rising when he woke, even if it happened to be as early as three o'clock in the morning. In 1679, he was appointed chaplain and almoner to the Princess of Orange, at The Hague in succession to his friend George Hooper, who had fallen out with William of Orange. In time, Ken also found himself unpopular with William. Sidney noted in his journal on 21 March 1680: 'Dr. Ken was with me, he is horribly unsatisfied with the Prince of Orange. He thinks he is not kind to his wife, and is determined to speak to him about it, even if he kicks him out of doors.'[34] In 1680, Ken returned to England, where Charles II appointed him a royal chaplain. Paradoxically he seemed to rise in the King's esteem when, on a royal visit to Winchester, Ken refused to countenance Nell Gwyn staying in his prebendal house in the city. Later, Charles was reputed to have said 'I must go and hear Ken tell me of my faults'.[35] When the diocese of Bath and Wells was vacant, Charles was said to have asked 'Where is the good little man who refused the lodging to poor Nell?'[36] Ken wrote to his friend George Hooper of his appointment:

> Amongst the herdsmen, I, a common swain,
> Lived, pleased with my low dwelling on the plain;
> Till up, like Amos, on a sudden caught,
> I to the pastoral chair was trembling brought.

Ken was consecrated by Sancroft and Turner on 25 January 1685, just before the death of Charles II.[37] A few days later he was called to the King's deathbed and spent some time with the dying Charles. Ken also insisted that the Duchess of Portsmouth, Charles's mistress, should leave the room and Queen Catherine of Braganza be admitted. But, like Sancroft, he was unsuccessful in getting Charles to receive Anglican communion, and, after Ken had left, the King was received into the Roman Catholic Church.[38]

William Lloyd was the only bishop of the seven to establish a reputation as a fierce anti-Catholic before his consecration. In 1676, he published *Considerations Touching their True Way to Suppress Popery*. The aim of the book was to distinguish between 'English Church Catholics' and Roman Catholics. Nevertheless James, Duke of York, said that 'Dr. Lloyd is a learned and worthy man, and has lately written a very excellent book'. Lloyd was appointed chaplain to Princess Mary on her marriage to William of Orange – through the influence of her former tutor, Bishop Compton of London. Lloyd, with Ken, was credited with ensuring that Mary remained an Anglican.[39] During the 'Popish plot', Lloyd was taken in by the fantastic claims, and fanned the flames by his funeral sermon for the murdered magistrate Sir Edmund Berry Godfrey, a supposed victim of the Roman Catholics. The sermon was delivered to an overflowing congregation in St Martin in the Fields with two able-bodied parsons standing guard on either side of his pulpit, as his personal defenders against the Popish forces. The peroration of the sermon was an appeal to keep England from the Catholic 'bloody religion'.

In the wake of the hysteria generated by the Popish Plot, Lloyd seems to have been an agent for Charles II, suppressing elements in the confession of one of the men framed for the murder of Godfrey. He earned rewards in the form of the vicarage of Llanfawr and a prebend of Llandaff. Lloyd was also commissioned with some secret work against subversive forces in Reading. His reward, or perhaps the King's desire to dispense with his services, came with his preferment to the diocese of St Asaph in October 1680. Despite his royal service, Lloyd was the only one of the seven bishops suspected of sympathy for Whig exclusionists.

Trelawny alone of the seven bishops was nominated to the bench of bishops by James. When James succeeded to the throne, he remembered the son of his old friend and household official, and appointed him to the bishopric of Bristol. Trelawny, never one to express much gratitude, wrote to the Earl of Rochester, Lord Treasurer, in July 1685:

> Give me leave to throw myself at your lordship's feet, humbly imploring your patronage, if not for the bishopric of Peterborough, at least for Chichester, if the Bishop of Exeter cannot be obliged to accept of that now vacant see, which he seemed to incline to when his removal to Peterborough was proposed; and I am assured from those about him, that if the King should be pleased to tell him he is resolved on his translation to Chichester, he will readily close

with it; and let me beseech your lordship to fix him there, and to advance your creature to Exeter, where I can serve the King and your lordship. I hear his Majesty designed me for Bristol, which I should not decline was I not already under such pressure by my father's debts, as must necessarily break my estate to pieces if I find no better prop than the income of Bristol, not greater than 300l. per annum; and the expense in consecration, first fruits, and settlement, will require 2000l. If Peterborough and Chichester shall be both refused me, I shall not deny Bristol, though my ruin goes with it, if it be the King's pleasure, or any way for his Majesty's service that I should accept it.[40]

Trelawny soon earned his mitre by active response to the Monmouth emergency. When Monmouth invaded the West Country, Trelawny mobilised the Cornish militia for the King. He was consecrated Bishop of Bristol, on 9 November at Lambeth Chapel, by Sancroft.[41]

To Sancroft's surprise, Trelawny showed himself, as a bishop, something of a stern Protestant. He wrote of his diocese to the Archbishop:

> The chiefest neglects which I found were the backwardness of people to be confirmed, occasioned by the neglect of constantly instructing the children in the words and meaning of the Church catechism; the ill custom of private christenings, through the minister's compliance with the richer sort of their parish; the disuse of visiting the sick at their houses, proceeding chiefly from the custom, which is very frequent, of reading most part of the form of the visitation of the sick when they are prayed for in the church; the confused and irregular way of reading the prayers, in some ministers, either through their own dissatisfaction at them, or fear of others dissatisfied with them; and the ill condition which most of the churches were in, by reason the parishes are not put in mind, or else unwilling to assess themselves for their reparation. The ministers, of whose faults in their disordered reading and praying I could make myself acquainted from good hands, I have taken care to punish, and I hope to their amendment.[42]

This reassured Sancroft, Trelawny was a conscientious bishop and not the layman in canonical dress the Archbishop feared. Trelawny also conducted a visitation in Bristol and Dorset, during which he sought to implement the royal policy of dampening down anti-Catholic preaching.

Preoccupations

Before James II's accession, the seven bishops demonstrated a number of interests and preoccupations. Sancroft was undoubtedly uneasy at the prospect of the succession of a Catholic. Soon after his consecration, in February 1678, Sancroft tried to persuade the Duke to return to the Church of England. Sancroft asked George Morley, Bishop of Winchester, who was respected by the King and the Duke of York, having attended Charles I during his imprisonment and followed the royal family into exile, to go with him to a meeting with the Duke. Sancroft took the lead however, and addressed the Duke with words which were subsequently widely published:

> What we are now about to say to your Highness is that which Heaven and earth have long expected from us that we should say, and what we cannot answer it to God or man if we omit or neglect, when we have an opportunity, which your Royal Highness is pleased at this time to afford us... You were born within [the Church's] then happy pale and communion, and were baptized into her holy faith. You sucked the first principles of Christianity from her, the principles of the oracles of God, that sincere milk of the Word, not adulterated with heterogeneous or foreign mixtures of any kind. Your royal father, that blessed martyr of ever glorious memory, who loved her, and knew how to value her, and lost his all in this world for love of her, even his life, too, bequeathed you to her at the last... When he was ready to turn his back upon an impious and ungrateful world, and had nothing else left him but this excellent religion (which he thought not only worth his three kingdoms, but ten thousand worlds), he gave that to the queen in legacy amongst you. For thus he spake to the King, your brother, and in him all that were his: 'If you never see my face again, I require and entreat you, as your father and as your King, that you never suffer your heart to receive the least check or disaffection from the true religion established in the Church of England. I tell you that I have tried it, and after much search and many disputes, have concluded it to be the best in the world...' You stood, as it was meet, next to the throne, the eldest son of this now despised Church, and in capacity to become one day the nursing father of it; and we said in our hearts, it may so come to pass that under his shadow, also, we shall sit down and be safe. But alas! it was not long before you withdrew yourself by degrees from thence (we know not how, nor why; God knows); and though

we were loath at first to believe our fears, yet they proved at last too mighty for us; and when our eyes failed with looking up for you in that house of our God, and we found you not, instead of fear, sorrow filled our hearts, and we mourn your absence ever since, and cannot be comforted. Now, you stab every one of us to the heart. Now, you even break our hearts, when we observe (as all the world doth) that we no sooner address ourselves to Heaven for a blessing upon the public counsels (in which you have yourself, too, so great and high a concern), but immediately you turn your back upon us. We pray, for your Royal Highness by name, and can you find it in your heart, sir, a heart so noble and generous, so courteous, too, to throw back all these prayers, and renounce them as so many affronts and injuries to Heaven and to you. If we who now stand before you, sir, should declare (as we do at present, and we hope it misbecomes us not) that we do now actually lift up our hearts, with our hands, unto God in the heavens, that he would be pleased to endue you with His holy spirit, to enrich you with His heavenly grace, to prosper you with all happiness, and to bring you to His everlasting kingdom; can you withhold your soul from going up together with our souls, one entire sacrifice to Heaven to so good and so holy a purpose? Or, if you can, which seems indeed to be the sad state of the case, nor is that action of yours (withdrawing from the prayers), in the common acceptation of mankind, capable of fairer construction, blessed God, what shall we say? It is more than time, sir, that you consider seriously between God and your own soul, when you two meet together alone at midnight, what you have done, and where you are; that you remember whence you are fallen, and repent; that at length you open your eyes; and we beseech Almighty God (who only can) to open your heart to better and more impartial information.[43]

The Duke replied that 'it was painful to be pressed on the subject of his religion just before the meeting of Parliament, as anything of that kind must increase the prejudices now prevailing against him'. Nevertheless, James did not harbour any resentment towards Sancroft and took his attempt to convert him as evidence of his regard for him.[44]

This meeting was significant for both men. For Sancroft it raised the issue of whether a Catholic king was entitled to the same high sacramental character as an Anglican. His attempt to convert James, at the very least, implied some discomfort at the notion. James, though aware of the exclusionist tendencies of some bishops, saw Sancroft as one of the Cavalier Tory Anglicans who could be trusted; indeed whose absolute loyalty could be taken for granted. Clearly neither man fully appreciated the viewpoint of the other.

Sancroft could not avoid becoming involved in politics. Charles dismissed the Cavalier Parliament in January 1679 and there were three elections in as many years. None of them gave the King the compliant Commons he sought to sanction James's succession. Sancroft naturally aligned himself with the King, and distrusted the Whigs, not least because he feared that their extremism was akin to that of the Parliamentarians of three decades earlier. Sancroft also saw the exclusion of James as a Whig attempt to make the monarch the creature of the Commons, as Cromwell had done. All this brought Sancroft into an alliance with the Hydes, the Earls of Clarendon and Rochester, James's brothers-in-law and the leaders of the 'reversionary interest'. Sancroft became a staunch supporter of the Duke: he attested to Monmouth's illegitimacy when he was mooted as a successor for Charles; he promoted the Duke's allies and supporters; he prosecuted Dissenters and commended James for his protection of the Episcopal Church in Scotland – which he felt boded well for James's succession. This marked Sancroft apart from Henry Compton, Bishop of London, who, though he was a High Churchman, regarded Catholicism with horror and foresightedly prepared for future possibilities.[45]

The King and Duke of York were suitably grateful to Sancroft and endorsed his Tory Anglican programme. The King issued a declaration in April 1680 affirming the laws on which the established Church and state rested. Sancroft was so delighted that he agreed for it to be read in every pulpit in England – a precedent that he may later have regretted. Sancroft also ushered in a period of prosecution of Dissenters in church courts, which was the ecclesiastical counterpart of the King's reinforcement of the Tory magistracy in corporations.[46] Together the King and Archbishop were building an alliance that would ensure the succession of James and, Sancroft hoped, secure the Church of England.

Mild and benevolent as he was in person – a feature that some mistook for weakness – Sancroft was nevertheless a strict and unbending enforcer of clerical discipline, and fiercely dealt with abuses. He suspended Thomas Wood, Bishop of Lichfield and Coventry, from the exercise of his episcopal functions, on account of neglect of his diocese even though Wood was under the patronage of the Duchess of Cleveland, the King's mistress. He also undertook metropolitan visitations of the dioceses of Lichfield and Coventry, Salisbury and Lincoln, which he felt to be inadequately administered by their bishops. These metropolitan visitations suggested that the Church was in a poor state and, while there were some who feared that James II was happy for the Church to be in such a low position, Sancroft was not. When the archdeacon of Lincoln was convicted of simony and petitioned the King for a pardon, Sancroft wrote to the King in a way that pointed to his future behaviour:

> Sire, the crime he stands convicted of, is a pestilence that walketh in darkness, too often committed but very seldom found out. And now there is a criminal detected, if your Majesty thinks fit, which God forbid, to rescue him from the penalty, the markets of Simon Magus will be more frequented than ever. Much rather, since he hath the courage to appeal to the delegates, to the delegates let him go...[47]

Sancroft also advanced high standards for admission to Holy Orders, issuing letters in 1678 and 1685 demanding tighter controls by the bishops. He also supported the Ecclesiastical Commission of 1681, which made ecclesiastical appointments, and used it to advance Tory churchmen. This was another decision he may have come to regret.

Sancroft remained committed to the divine right of kings and to an absolutist view of authority. At the time of Charles II's death, Sancroft had commissioned the Tory writer Edmund Bohun to revise Filmer's *Patriarcha*, the most strident advocate of divine origins for government and royal authority.[48] The succession of James was something of a triumph for Sancroft. Despite his failure to respond to Sancroft's urging to return to the Church, James had soothed the apprehensions of many people by a voluntary declaration at his accession council that it was his intention to protect the Church of England. The Archbishop responded with an expression of loyalty and assurance to James that the Church's 'holy boast, that she hath been always loyal to her kings'.[49] Mutually reassured, Sancroft performed the coronation of James on St George's day 1685.

As Beddard claimed, for most of James's reign Sancroft was left 'to pick a precarious middle way between compliance and truculence'.[50] Burnet claimed, rather unfairly, that Sancroft remained silent at Lambeth; in fact he was far from passive. He refused to order the clergy to abandon afternoon catechising as James wanted, he received the aggrieved dons of Oxford and Cambridge as James's incursions began there and he joined the governors of Charterhouse in refusing to admit a Catholic on the King's orders. Gradually Sancroft was finding his way from loyalty to opposition. When James issued his first Declaration of Indulgence in April 1687, he knew that the choice of supporting James or Dissent was now swinging in favour of Dissent, and Sancroft inaugurated discussions with Protestant Dissenters as to how they could unite against James's policy to defend the Church. It was no coincidence that, in October 1687, Sancroft authorised *The Speculum Considered*, which joined the debate, begun earlier in the year, in which Thomas Tenison had attacked Catholicism. *The Speculum Considered* refuted the claims of Catholicism and was a signal of Sancroft's thinking.

In October 1687, the Princess of Orange, recognising the importance of Sancroft in the events unfolding in England, wrote to him:

> Though I have not the advantage to know you, my Lord of Canterbury, yet the reputation you have makes me resolve not to lose this opportunity of making myself more known to you than I could have been yet. Dr. Stanley can assure you that I take more interest in what concerns the Church of England than [in] myself; and that one of the greatest satisfactions I can have is to hear how that all the clergy show themselves as firm to their religion as they have always been to their King; which makes me confident God will preserve His Church since He has so well provided it with able men.[51]

In the highly political events leading to the trial of the bishops and the Glorious Revolution, it was easy to ignore Sancroft's churchmanship. An important element in Sancroft's theology, which influenced his actions as much as his ecclesiology, was patristic scholarship and his advanced views of the sacerdotal nature of both the episcopate and kingship. As a biblical scholar, Sancroft was deeply immersed in the early church and its struggle against imperial authority. St Paul's injunction to the Romans 'Let every soul be subject unto the higher powers' was also an important element in his thinking.[52] In this, Sancroft was a characteristic High Church Caroline divine. But Sancroft also drew on elevated view of anointing, both of bishops and kings. Anointing was a biblical act which transmitted an indelible character to the consecrated. The biblical events of King David's reign offered a direct parallel to the events of 1685–8, and Charles II had been consciously identified as David throughout his reign, most notably by Dryden who wrote

> Thus banish'd David spent abroad his time
> When to be God's Anointed was his crime.[53]

King David designated Solomon as his heir and Zadok the priest anointed him King; yet Adonijah – another of David's sons – sought to displace Solomon, and was killed for his usurpation. In this analogy, Charles II/David designated as his heir James II/Solomon and was anointed by Sancroft/Zadok, but he was displaced by William/Adonijah. Once James had been anointed by Sancroft, his kingship was as fixed and immutable as Sancroft's own episcopal authority.[54] This explains why blandishments about oaths in 1689 did not change Sancroft's mind. James might have violated his coronation oath, he might have broken an original contract with his subjects, he might even have abdicated the throne, but he could not be unanointed or unkinged, any more than Sancroft could be unconsecrated. Similarly, Sancroft was willing to conspire and trim to confront James, and even allow William to restore the Church of England's privileges, but he could not remove the sacred oils from James's forehead.[55]

John Lake was also keen to show his loyalty to James. In the summer of 1685, he was despatched from London during Monmouth's rebellion; James considered the presence of a loyal bishop, like Lake, would be vital in keeping the city of Bristol quiet, and he asked Lake to go there as soon as possible. Lake, who was confined to his bed with an attack of gout and unable to move hand or foot, obeyed the King's command and travelled there by coach. He was in danger of falling into the hands of the rebels at Keynsham bridge, but eventually safely reached the city. He gave an account to Sancroft, written in July 1685, and did not neglect the opportunity to importune for a better diocese:

> At present we are celebrating, with all expressions of joy and triumph, the late happy victory obtained against the rebels, who are now totally dispersed; and this, accumulated with the news that the late Monmouth and Grey are taken in Dorsetshire, which is newly brought to the Duke of Beaufort; but of this he waiteth a confirmation. I durst take confidence, I would presume, with respect to myself, to tell your grace that I have been advertised this day that the Bishop of Chichester is dead, and if your grace could think fit to move for that bishopric for me, it would be more convenient than Peterborough; but in this, as in all other things, I must refer myself wholly to your grace.[56]

Sancroft recommended Lake to the grateful King, who, given Lake's loyalty, granted his request, and appointed Lake to the vacant see of Chichester. Lake's translation to the see of Chichester took place in

October 1685. Writing to Sancroft, Lake noted the neglect that had affected his new diocese:

> conventicles are set up in most of the great towns in the diocese, and ... papists are very busy to make proselytes, but with little success. They have indeed gained four or five, in one parish, but they are poor, mean persons, who bring neither credit nor advantage to their church; nor do I hear of so many in the whole diocese beside, and yet divers of the most considerable persons in our part of the country are zealous that way.[57]

During the three years he held the see of Chichester, Lake further demonstrated his High Church credentials by establishing weekly communion in the Cathedral. But he also restored the sermons to the nave of the cathedral, which was a move to placate Dissenters; consequently Lake drew a number of Dissenters to the Church of England. These policies shed some light onto Lake's attitudes and perhaps gave the lie to James's assumption that Tory Anglicans were irredeemably hostile to Dissent. Lake's wooing of Dissenters shows that just as Tory Anglicans wanted to rebuild the physical structures of the Church, they also sought to rebuild it doctrinally. In this, Lake was close to Compton's belief in the unity of Protestants. Lake, lacking the private income of Sancroft, was of necessity a place-seeker, but under James he had prospered, and could perhaps have expected further rewards.[58]

During the Monmouth emergency, Thomas Ken also responded to the King's call and went to Wells where he worked hard for the relief of the rebel prisoners, visiting those who were imprisoned in Wells, and paying for their food. James recognised that Ken's actions 'proceeded not from disaffection to his person or government, but from motives of compassion to so many distressed brethren whom he saw in danger of perishing both soul and body, thanked him for what he had done'.[59] As a proof of his full confidence in the loyalty of Ken, James appointed him to attend Monmouth at his execution.

In the winter of 1685, Ken delivered a series of lectures on the catechism in the chapel of Ely House in London. Ken's eloquence attracted large crowds and the lectures were afterwards published, and widely admired. Even Princess Anne went privately to hear Ken, and requested that some place might be provided where she might do so without being recognised.[60]

Ken was placed in a difficult situation by James's visit to Bath in 1687. The King indicated 'that he intended to exercise the royal gift of touching, for the cure of the evil, in the abbey on Sunday, after morning prayer'. He had already touched for the evil in Chester.[61] Ken, who was at Wells, was annoyed and perplexed when told of what was about to happen in Bath Abbey; but he had no means of stopping it without provoking a confrontation. Ken was clearly ambivalent about the matter, and he wrote to Sancroft:

> When his Majesty was at Bath there was a great healing; and without any warning, unless by a flying report, the Office was performed in the church between the hours of prayer. I had not time to remonstrate, and if I had done so it would have had no effect but only to provoke; besides I found it had been in other churches before, and I know of no place but the church capable to receive so great a multitude as came for cure, upon which consideration I was wholly passive. But being well aware what advantage the Romanists take from the least seeming compliances, I took occasion, on Sunday, after the Gospel, the subject of which was the Samaritan, to discourse of charity; which I said ought to be the religion of the whole world, wherein Samaritan and Jew were to agree; and though we could not open the church doors to a worship different from that we paid to God, yet we should always set them open to a common work of charity, because in performing mutual offices of charity one to another there ought to be an universal agreement.[62]

Ken's offence at James's healing in his diocese, and his concern for the unity of Protestants, gave him pause for thought, and his concerns intensified during James's reign. In April 1688, Ken issued a pastoral letter. It was not a normal practice for bishops in the diocese to do so, and therefore the letter attracted some attention. It was difficult for the letter not to be viewed as a coded expression of concern at the King's policies. Ken urged the clergy to spend Lent considering their own sins and 'the sins of the nation'. He pointedly asked them to give bread to the hungry and poor, especially 'the many poor Protestant strangers...now fled thither for sanctuary' from James's Catholic ally, Louis XIV. Ken also asked the clergy to treat the Church as a bulwark for Protestantism: 'your greatest zeal must be spent for the Publick Prayers in the constant devout use of which, the Public Safety both of Church and State is highly concern'd.' He sought to reduce the numbers of Dissenters from the Church, at a time when the King was also wooing Dissenters and aimed to 'promote universal charity to all that Dissent from you'. He asked the clergy to go from house to house warning everyone 'night and day' against the dangers of 'public provocations' and 'national guilt'. Ken's pastoral letter was a fairly thinly disguised comment on the issues of the moment. Gone was the episcopal antagonism towards Dissenters that his predecessors at Bath and Wells had pursued. In its place was a sense of the unity of Protestants against the growing national problem of James's policies. For such a pastoral letter to be issued in Bath and Wells, the cockpit of radical Protestantism, seemed like an encouragement to those forces which opposed James.[63] G. H. Jones has suggested that Ken only cooperated as long as he did with the King's policies because he feared schism.[64] But, as has been seen, Ken knew William and therefore realised more than most that the alternative to James was not quite as rosy as some assumed. It also demonstrated the growing equivocation of Ken to James's policies, despite his dislike of William of Orange.

As if to press home the issue even more explicitly, Ken issued a second letter to the clergy of his diocese on 14 April 1688 on the issue of French Protestants. He made much of the fact that God had put into the King's heart the desire to grant protection to them, allowing them to seek alms and charity. Ken emphasised that it was the duty of his clergy and diocese to 'do good to all men, especially to those who are of the Household of Faith'. But Ken's emphasis was on brother Protestant rather than all Christians, which would include Catholics. Once again, Ken had launched a flare signal to his diocese and one which they would not have misunderstood.[65]

Ken's reaction to James's policy was also seen at Easter 1688. He was appointed to preach the afternoon sermon in the Chapel Royal at Whitehall, on Passion Sunday, in April 1688. The morning sermon had been preached by Edward Stillingfleet and holy communion followed, but was interrupted by the mob breaking in, eager to hear the sermon to be preached in the afternoon by Ken. John Evelyn recorded 'In the morning service the latter part of that Holy Office could hardly be heard, or the sacred elements be distributed without great trouble.' Crowds fought to secure seats, so that the chapel was already overflowing before the arrival of Princess Anne, who took her place in the royal gallery. When prayers were over, the bishop ascended the pulpit, and read his text from Micah, ch. 7, v. 8–10, it was an especially apt one:

> Rejoice not against me, O mine enemy; when I fall I shall arise; when I sit in darkness the Lord shall be a light unto me. I will bear the indignation of the Lord because I have sinned against Him, until He

plead my cause and execute justice for me. He will bring me forth to the light, and I shall behold His righteousness.

Ken represented the Church of England as Judah, the Roman Catholics as Babylonians and the Dissenters as the Edomites. He lamented that 'he had not, like Micah, the happiness of having the King himself for an auditor; therefore his discourse might possibly be misrepresented to him, since the very Scripture itself might be perverted by insidious men'. The next Sunday, Ken preached at St Martin in the Fields and Princess Anne again attended to hear her father's religion denounced and the Dissenters called on to aid the Church.

Reports of Ken's sermons were made to the King, who was furious and sent for him. James expressed both surprise and displeasure at hearing that seditious doctrines had been preached from his pulpit in the Chapel Royal at Whitehall. Ken replied 'If your Majesty had been happily present in your proper place, mine enemies would not have had the opportunity of bringing a false accusation against me.' This reproach put the King further out of temper, and he dismissed Ken without another word.[66]

Unlike the other bishops, Trelawny had secular interests, he was patron of a number of Cornish boroughs, and reported intelligence to Sancroft when, in 1687, it appeared that there might be an election. Such information was important to both Sancroft and Trelawny in measuring reaction to James's policies:

> We have had frequent alarms that a parliament is speedily intended, to which Cornwall sends forty-four [members]; and knowing myself to have a good interest in the gentry, I was resolved to see what inclinations they had, and what courage to support them in case of an attack from the lord lieutenant; and I was glad to find the gentry unanimous for the preserving the Test and our laws; and what pleased me much, resolved to appear in their several corporations, and not suffer so many foreigners to be put upon them, as were returned hence by the wheedle of the Earl of Bath, the lord lieutenant, whom now they will attend in a body upon his coming into the country, and, with the decency of a compliment, desire that they themselves may be permitted to serve the King in parliament; which, if his lordship will not yield to, but answer that he has the King's command for the return of such as his Majesty named to him, the gentry, at least a great part of them, will attest their particular intentions in such boroughs as have dependences upon them,

and try whether the Earl of Bath will, with a high hand, turn out such mayors and magistrates as will not comply with his nominations, disoblige the gentry, and endanger the kingdom. I have one thing more to acquaint your grace; that being with several of our gentry when the news came that the several lord-lieutenants should call together their respective deputies and the justices of their counties, and know of them whether they would or would not take off the Test; my opinion was that they should not give a plain answer whether they would or would not, but only in general, that if they were chosen they would be governed by conscience and reason; for should they say downright they would not take off the laws and the Test, there would be a positive command that all such as had declared themselves of that opinion should not be chosen.[67]

This intelligence was vital, it encouraged Sancroft in his opposition to the King and made him see what a useful ally he had in Trelawny. For Trelawny, the role of civil as well as ecclesiastical magnate gave him an insight into the way corporations and laymen feared James's Catholicism. Before the start of 1688, Trelawny had become an adherent and secret correspondent of the Orange faction. Nevertheless, James regarded him as a close ally. Hence, when Trelawny signed the petition, the King called him 'the most saucy' of the bishops.[68]

Sceptical loyalism

The seven bishops were undoubtedly taken by James to be in the vanguard of his loyalists. In their resistance to James, they were, as G. M. Trevelyan wrote, 'milder and more conservative than the five members who Charles I attempted to arrest'.[69] They mostly came from loyalist Tory cavalier Anglican families and were Stuart loyalists. Sancroft and Lloyd had even visited the court in exile. If James needed evidence of their political affiliation after the Restoration, he would have been heartened by their Toryism. They had backed him during the exclusion crisis and Turner had even suffered internal exile in Scotland with James. Perhaps James was also conscious that they were, for the most part, men who were willing to supplicate for preferment. Lake, Lloyd and Trelawny were clearly 'clients' of James; in this respect they owed him an obligation, and obligation implied a return. If James needed evidence of the practical help the bishops could afford him in defending his kingdom, there were few better examples than the responses of Bishops Ken, Trelawny and Turner during the Monmouth

Rebellion. They returned to their dioceses and risked their lives to defend the regime against the Protestant Duke and his rebels.

Sancroft, Lake and the others had an elevated view of the sacraments – hence they tended to advocate frequent celebrations of holy communion. This sacramental view also applied to kingship, but kingship owed its sacramental character to the anointing by the bishop at the coronation. This was an Anglican anointing and a sacramental kingship conferred by Anglican bishops, who traced their succession back to the apostles. However, James's assumption that consequently the bishops would endure all his policies without complaint and without resistance was profoundly flawed. And the signs were there for those that looked for them. Even High Church Tories, like Edmund Bohun, were clear that the doctrine of passive obedience did not mean colluding in misgovernment or subverting the establishment of Church and State. Nor did passive obedience mean 'abetting a prince to enslave a people'.[70]

A key element in the bishops' willingness to confront James was Sancroft's character and leadership. Sancroft was an ascetic and a celibate who had shown during the Commonwealth that he was happy living an almost solitary life in Norfolk. His interest in Charles I's martyrdom and his dislike of political intrigue made him more likely to seek obscurity than to cling to power as a matter of expediency. Moreover Sancroft – and William Lloyd – were convinced that the Book of Revelation's prophecies were 'fulfilled in contemporary events'.[71] By 1688, there was, consequently, a fatalism and passive acceptance of what Sancroft saw as inevitable. Sancroft was also a builder, being partly responsible for the new St Paul's Cathedral, and he also endowed other churches and buildings. He shared this role with Lake and Turner. But Sancroft, Lake and Turner did not just rebuild the temporalities of the Church, they had reconstructed its spiritualities too.

Sancroft and the other bishops conceived and fashioned the Restoration Church of England as a national church with moral and spiritual authority over all men and women in the country. This was the aim of Cranmer as much as Sancroft. Thus Sancroft would not tolerate incapacity in the bishops of Salisbury, Lincoln, and Lichfield and Coventry, where he conducted metropolitan visitations. He also sought energetic and able churchmen for his parishes and dioceses, and was intolerant of corruption and ineptitude. Sancroft's views of the Church are what led to his unease at Catholic army officers and James's inhibition of consistory court actions against Catholics. Sancroft had a high view of the apostolic succession, which had placed him in the chair of St Augustine. It was James's usurpation of Sancroft's right to dispense discipline to his suffragan bishops, like Henry Compton, and the King's investment of that right in the Ecclesiastical Commission which opened the breach between Sancroft and the King.

Like Sancroft, Ken's saintly exterior disguised a steely and determined will. His ambivalence towards James's Catholicism and his touching for the 'King's Evil' were modest concerns in comparison to Ken's sense of the unity of all Protestants. Ken's defence of Huguenot refugees, his insinuating pastoral letter and Easter sermon of 1688 all showed that Ken owed more allegiance to the Church than to the King. The same allegiance was shared by the bishops and many people in England.

The seven bishops had shown themselves sympathetic to James during the Whig Exclusion crisis, but, significantly, Sancroft Turner and Lloyd had been taken in by the extravagant and implausible claims and allegations of Titus Oates during the Popish Plot. They had been ready to believe the worst of Catholics and to give credence to ideas of a Catholic conspiracy. Turner and Lloyd were stern in their misgivings about Catholicism, even after the discovery of the fraud that Oates had perpetrated. Both preached sternly against Catholicism – to the King's annoyance.

Of all the bishops, Lloyd and Trelawny were the closest to being clerical politicians. Like Compton, both had established contact with William before the confrontation with James and, by 1688, both – unlike Sancroft and Turner – were realists in recognising that James had to go. A regency or other compromise was not good enough. What Trelawny possessed – through his status as the scion of a county family – was a deep understanding of what the laity felt and thought. Once James had turned away from the Tory landed classes in 1687–8, Trelawny had further cause to question his loyalty to James. When Trelawny reported to Sancroft that the lords lieutenant and gentry of the West Country would not tolerate an end to the Test Act, he was reporting what the other bishops knew from their archdeacons and clergy. Though at first glance in 1685 James had little to fear from the seven bishops, there were already bat-squeaks of discomfort and concern. What James did not understand was that each of the bishops, when forced to choose between Church and King, would always choose the Church.

2
The King's Policies 1685–7

Early days and Monmouth's rebellion

Many of James's problems were those that faced his brother. Patrick Dillon claimed that James 'inherited a slow-burning crisis whose fuse was still alight'.[1] It was a fuse on which James was determined to blow. By the end of his reign, Charles II had thrown in his lot with the Tory Anglicans. He regarded Whigs and Dissenters as dangerous and potentially treasonable. This is what Michael Mullett called the 'Second Restoration' which, in the wake of the Rye House Plot, saw the Tories ascendant and the Whig exclusionists at bay.[2] But James failed to recognise how much Charles had become dependent on the Tory Anglicans.[3] The consequence was that both James and the Tory Anglicans failed to recognise the slow-burning nature of the problems that faced them, and in particular the unrealistic expectations each had of the other.

James's reign began in mistrust. Despite the new King's welcome declaration at his accession council, in which he denied he was an adherent of arbitrary power and promised to 'maintain the government in Church and State as established by law', Gilbert Burnet claimed that James privately regarded the Elizabethan settlement of the Church as illegal and therefore he had little or no faith in the King's words. This distrust was encapsulated in the council's request that James publish his promise to defend the Church. On the following day, James told Sancroft and Turner that 'he would never give any sort of countenance to Dissenters knowing it must needs be faction and not religion if men could not be content' with the current establishment.[4] However, James's dismissal of Lord Halifax, as lord privy seal and a member of the Privy Council, in February 1685 added to the sense that James was not a King who would heal wounds. Soon afterwards, Burnet left for the continent.[5]

There were others who were wary of James. Bishop Henry Compton of London, fearing that the effusive addresses of congratulations from the dioceses would encourage James's autocratic tendencies, ensured that the London diocesan address included an explicit reference to 'our religion established by law, dearer to us than our lives'.[6] Some Tories, like Halifax and Nottingham, reacted cautiously to James though they were prepared to give him the benefit of the doubt at every turn.[7] But the moderate Tories were divided, Danby having animosity for Halifax. Moreover Danby was relieved of the threat of impeachment for his role as lord treasurer under Charles II, but only because James wished to expunge the impeachment of three Catholic peers at the same time.[8] Nevertheless, Anglicans also felt obliged to give James the benefit of the doubt; John Sharp, later a thorn in the King's side, was among those who thanked James for his declaration in support of the Church.[9]

If some bishops and politicians distrusted James, there was an equal and opposite sense of unease. Archbishop Sancroft had an early warning of the direction James's mind was taking, in March 1685 James told the Primate:

> I will keep my word and undertake nothing against the religion which is established by law, unless you first break your word to me. But if you do not do your duty towards me, do not expect that I shall protect you. You may be sure I shall find means to do my business without you.[10]

Why would James make such a statement? Clearly James had some 'business' in mind, and it was likely that this was the repeal of the Test Act, which buttressed the religious establishment. Early in his reign James also ordered Sancroft and the other bishops to forbid the clergy from preaching seditious sermons – and he made this under threat of removing his promise to protect the Church. By seditious sermons, James clearly meant those which attacked Catholicism. Sancroft promised to reissue the 1662 directions to preachers that they should avoid speculative and controversial issues.[11] It was widely rumoured that royal spies were placed in London churches to report which clergy observed the injunctions regarding preaching.[12]

James's coronation demonstrated the ambivalence of the nation. Only about half the peers qualified to attend came to the ceremony.[13]

James shocked many by holding a Catholic mass after his crowning, and Francis Turner preached an extraordinary sermon which was an important signal to the King. The sermon grappled with a problem that affected many Anglicans: they did not believe their obedience was conditional because they expected James would keep his word and coronation oath. But there were those who believed that James might abuse his power and this damaged the nature of passive obedience to the King.[14] Turner chose as the text for his sermon the coronation of King Solomon from the Book of Chronicles – a motif strongly in Sancroft's mind as has been seen. In many respects it was an appropriate text for the event, though Turner mishandled it. Turner compared James's crowning with Solomon's coronation for three reasons. First, Solomon's title was 'firm and good'; second, his government 'was as good as his title' and 'his people were an obedient people'. But Turner's sermon struck some discordant notes. In praising the Queen, Turner mentioned that she had shared 'her royal husband's sufferings and hard travels' – a reference to James's exile during the Exclusion crisis. In claiming that James's accession commanded popular support, Turner felt obliged to deny that 'we imagine our united voices contribute anything of right in our hereditary prince'. Rather than leave aside the issue of James's title to the throne, Turner pressed on to assert that no usurper could expect to 'reign prosperously' and that any questioning of James's claim was dangerous 'else there will be competitors'. Turner went on with remarkable similes, he asserted that 'management of the sceptre' had to be as strong as the King's claim, and pointed to the precedents of 'the second Edward and Richard' as kings who had indisputable claims to their thrones but lost them through misgovernment. The second James may well have felt these were entirely infelicitous precedents.

When Turner came to the loyalty of subjects, he plunged blithely on. He came close to contract theory when he claimed, 'since the wills of men are free, tis confest their leaves must be asked, whether they be happy or no; whether they will obey...For want of a people obedient and willing to be ruled by a gentle hand, the best of kings was most vilely cast away'. But he also argued that people ceased to be good and religious when they rebelled. Turner then turned to the issue of James's claim and said that, having been at Charles II's deathbed, he could attest that the King had wanted to be succeeded by his brother. He spoke of the deliverance 'from that abominable Excluding Bill...' and warned those who would challenge James: 'take heed of destroying your country to build your own house'. At the end of this badly misjudged sermon, despite his intentions, Turner's hearers must have felt that James's title to the throne was questionable, his peace dependent on his subjects' compliance and his own success dependent on his wise rule. This cannot have been James's intention in choosing his former chaplain to preach his coronation sermon.[15]

Initially James permitted the prosecutions of Dissenters under the Test Act and the Clarendon code to continue as they had under the Tory Anglican alliance of Charles II. Richard Baxter, for example, was convicted in May 1685 of scandalous and seditious libel for publishing his *Paraphrase on the New Testament*.[16] At the same time, James stopped all legal actions against Catholics who had suffered for their loyalty during the Civil War.[17] Thus the Tory Anglican agenda seemed to continue largely unaffected by the succession. James's experience of having to leave England because of the Exclusion crisis engineered, as he saw it, by Whigs and Dissenters as a means to prevent his succession and to diminish his brother's royal powers, was one which made him bitter towards his political opponents. He therefore sanctioned the persecution of Dissenters 'on political grounds'.[18]

The elections to the Parliament following the coronation indicated the high hopes that the country had of James. Even Compton asked his rural deans to promote candidates who would act for the King's service. The avowed Whigs in Parliament fell to just 30 or 40, and were badly demoralised.[19] But Compton's doubts soon returned, especially when James attended the celebration of mass publicly, despite the law against it. When Parliament met, it was dominated by Tories, and it granted James a generous civil list and supply of revenue. But in May 1685 a committee of the Commons petitioned the King to enforce all the laws against religious nonconformity, and the King reacted with anger. In the event, the Commons withdrew its petition and, instead, passed a motion expressing its satisfaction at the King's pledges to defend the Church of England. Later a bill was introduced to include in the definition of treason incitement of hatred of the King, and the Commons added a rider that it was lawful to defend the Church of England against Catholicism. James refused to proceed with the bill, and it lapsed.[20] This was the most favourable Parliament James could have obtained and yet it refused to be compliant on the issue of religion.

Early in his reign, James was determined to demonstrate that the Popish Plot had been a fraud. He had been outraged at the persecution of Catholics during the hysteria of the plot. Hundreds of James's co-religionists had been jailed and 24 were executed. James regarded

Sancroft and Lloyd with some suspicion because of their gullibility during the plot and, in the case of the latter, suspected collusion in it. The Popish Plot was also linked to wild allegations of attempts to assassinate Charles II and, at one point, James himself had been accused of involvement in it. The residual bad feeling was largely seen in the stories that abounded: it was said Princess Anne believed her father might have poisoned Charles II in order to succeed. But it gradually dawned on most people that the plot was a fraud to discredit the Catholics, and participants in the plot were embarrassed at their credulousness. Lloyd wrote to Sancroft denying that he had ever believed the confessions extracted during it; though Lloyd had attended the executions of some of the plotters. But James turned a deaf ear to Lloyd's hints that he would like to succeed to a better see than St Asaph when, in turn, the bishops of Chester, Oxford and later York were ailing or dead.[21]

The summer of 1685 was dominated by the emergency of the Monmouth Rebellion. The Duke, Charles II's illegitimate Protestant son, landed in the West and, after a foray as far north as Bath, faced the King's army at Sedgemoor. The rising caused the Tories and Anglicans to close ranks with James. Bishop Mews, for example, lent the King's army his carriage horses to haul the royal cannon at Sedgemoor and John Churchill commanded one of the royal regiments. James used the emergency to commission 86 Catholic officers in violation of the Test Act. The speed with which the Monmouth Rebellion was crushed gave James a new sense of confidence, and providential protection. During the emergency, James had also been heartened by William of Orange's willingness to send English troops stationed in Holland to use against the rebels. This despite the snub James had administered to William by refusing to grant him the style of 'His Royal Highness'. James was also encouraged by the fact that moderate Tories, like Danby, refused to have anything to do with Monmouth.[22]

The clergy who attended Monmouth at his execution, including Tenison, who was known to be sternly anti-Catholic, exhorted the rebel Duke to accept that the Church of England taught the detestation of rebellion. Certainly Tenison had not reached a point at which he would countenance rebellion.[23] Consequently, James felt he could still rely on the obedience of the Church. James told the French ambassador, Barillon, 'I know the English; you must not show them any fear in the beginning'. But Barillon was left with the feeling that James regarded any disagreement as tantamount to rebellion and placed too much faith in displays of power.

After the brutal repression of the Monmouth Rebellion, John Churchill, despite being active in the King's service, remarked that James had the heart of a slab of marble. In the wake of the rising, James took the opportunity to keep his standing army at 20,000 soldiers. He permitted Catholics to continue to hold commissions and also felt strong enough to ban the 5 November celebrations (which he felt were too anti-Catholic in tone) in 1685. For some moderate Tories, these were the first causes of concern. Humphrey Prideaux, writing from Oxford, commented that 'we have now got a standing army, a thing the nation hath long been jealous of; but I hope ye King will noe otherwise use it than to secure out peace'.[24] Concern at a standing army certainly brought about a rapprochement between Halifax and Danby, Halifax having spoken against the Catholic army commissions in Parliament along with Bishop Compton.[25] In the Commons, Sir Thomas Clarges said what few others had the courage to voice: that James's actions raised the prospects of a Popish army.[26] James also dismissed Henry Sidney, the commander of the English troops stationed in Holland, since Sidney was a Whig who had supported the Exclusionists in the late 1670s. James told the other members of the Privy Council 'that he would place no confidence in anyone whose principles and sentiments were opposed to his'.[27]

In France, James found a model of kingship which he hoped to emulate. On 2 October 1685, Louis XIV revoked the Edict of Nantes, removing religious tolerance from his Protestant subjects. Louis's commander in the principality of Orange, William's own patrimony, imprisoned pastors, burnt Protestant Bibles and forcibly converted the remaining population to Catholicism. In the view of Gilbert Burnet this was 'the fifth great crisis of the Protestant religion'.[28] In November 1685, the Catholic Bishop of Valence urged both James and Louis to remove heresy – by which he meant Protestantism – in their realms. This coincided with James's decision to enlarge his army.[29] Anglicans were sympathetic to the refugees from France, and bishops and clergy accommodated and raised funds for them. More worrying, was the fear among Whigs that the way Louis ruled France was a model for James. Evidence of the fierce punishments meted out to Protestants in England was exemplified by the prosecution of Dr Samuel Johnson, rector of Corringham, Essex, who published an anti-Catholic tract entitled *An Humble and Hearty Address to all English Protestants in the Army*, and, more dangerously, *The Opinion is this, that Resistance may be used, in Case our Rites and Priviledges shall be invaded*. Johnson was sentenced to be degraded from the priesthood, stand three times in the pillory, fined

500 marks and whipped from Newgate to Tyburn (which earned him 317 stripes in total).[30]

Compton and the Ecclesiastical Commission

Henry Compton was a source of considerable irritation to James. Compton's epithet of 'the Protestant bishop' had been earned during Charles's reign by virtue of his conversations with Dissenters in an attempt to reunite them with the Church. Compton was also an aristocrat, son of the Earl of Northampton, and well-connected with the court in The Hague, having been tutor to Mary, and was still in contact with her via his friend William Stanley, who was Mary's chaplain. Compton also presided over an important diocese, in which he protected and defended the right of the clergy to speak their minds from their pulpits, especially when it came to condemnation of Catholicism. He gave London livings and encouragement to preach his brand of Protestantism to some of the most important preachers of the day including, Sharp, Sherlock, Clagett, Tenison, Fowler and Horneck. Ryle claimed that, between them, these anti-Catholic preachers poured over 20,000 pages from the press at this time.[31] Compton maintained a unity of purpose among his clergy by holding regular synods of the clergy in the diocese. Within nine months of his accession, James's tolerance for what he saw as Compton's contemptuous anti-Catholicism ran out. He called Sancroft to a meeting at which he ordered the Archbishop to suppress afternoon lectures in the diocese of London. James's ostensible reason was to permit more time for catechising, but both Sancroft and Compton realised that his real goal was to silence those preachers who spoke against Catholicism. Compton took heart from the fact that the lectures had been established by the Act of Uniformity of 1559 and could not therefore be silenced by the King's whim. Compton ignored the directive, simply issuing a letter to lecturers asking them to moderate their language.

Compton had also annoyed James in the House of Lords, in autumn 1685, by opposing the permanent commissioning of Catholic army officers as contrary to law. James archly said 'I am determined not to part with any servants on whose fidelity I can rely, and whose help I might perhaps soon need'. In the debate on the issue, Compton predicted that James's introduction of officers into the army was a prelude to their admission to higher posts. In a significant contribution, Compton said 'the laws of England were like the dykes of Holland, and universal Catholicism like the ocean – if the laws were once broken, inundation would soon follow'.[32] Following his lead, there were claims by other peers that in peacetime the only thing for an army to subdue was Magna Carta. An unread speech by Danby, drafted for the November parliamentary meeting, appealed to the King to seek the judges' opinion on whether he possessed a dispensing power which would enable him to grant the Catholic commissions.[33] The bench of bishops agreed with Compton. For this act of defiance, Compton was dismissed as dean of the chapel royal, and Bishop Crewe of Durham was appointed to replace him; Compton was also dismissed as clerk of the closet to which Sprat of Rochester succeeded. Two days before Christmas of 1685, Compton was dismissed from the Privy Council also. His former apartments in Whitehall Palace were granted to Father Petre, James's confessor, who now also attended the Privy Council meetings.[34]

In the same session of Parliament, James also proposed the repeal of the Test Acts. But in November 1685 and early in 1686, Parliament refused to consider both the repeal and the permanent commissioning of Catholic army officers, and therefore James dissolved it. He claimed that the militia had made a poor showing during the Monmouth Rebellion, and therefore it was necessary to expand the army and to confirm Catholic officers in their commissions. But the Commons refused taxation for 'additional forces'. Parliament also sent an address to James saying that the employment of Catholics was illegal; James's response was that he had not expected such an address.[35] The only reaction James could pursue was to use the royal prerogative to pardon 60 Catholic officers who were prosecuted for not taking the Test during their tenure in the army.[36]

James's conviction that his authority enabled him to dispense with the law to permit Catholics to hold office appeared to be endorsed by the judgement in the Hales case. Sir Edward Hales had served as an MP from 1661 to 1681. In November 1673, he was appointed colonel of a regiment of foot and later served as one of the lords of the Admiralty. He was a trusted friend of James II, who appointed him a Privy Councillor, deputy-governor of the Cinque Ports, lieutenant of Dover Castle, lieutenant of the Tower of London and master of the Ordnance. In November 1685, Hales was received into the Catholic Church. James confirmed his command of a regiment of foot, for which the Test Act required him to take the oaths of supremacy and allegiance, receive holy communion in the Church of England, and make a declaration against the doctrine of transubstantiation; all of these actions would be impossible for a Catholic. Hales's servant, Arthur Godden, acting on instructions from Hales who wanted to challenge the law, brought a legal action against his employer. In March 1686, Hales was convicted at the Rochester assizes of failure to meet the requirements of the Test Act. Hales then appealed against his conviction to the King's Bench. He claimed that letters of patent from the King allowed him to hold his commission without taking the required oaths. The case was really about whether the King had the power to grant dispensations from religious penal laws in individual cases. The whole case was orchestrated by the government to prove this legal point. The appeal was heard by Lord Chief Justice Herbert and eleven other judges. In his summation of the case, Herbert compared the right of the King to suspend laws to that of God, who suspended his own laws in his command to Abraham to kill his son.[37] By a majority of eleven to one, the judges found in favour of Hales, affirming the King's dispensing power 'in particular cases and upon particular necessary reasons'. Some saw this phrase as a clear limitation of the King's dispensing power and not an affirmation that he had a general dispensing power. It seems likely that James either did not appreciate the difference between a particular and general power, or did not wish to. Indeed James cited the Hales case when he admitted Catholic peers (Powis, Arundel, Balasyse, Dover and Tyrconnell) to the Privy Council in the autumn of 1686.[38] Modern scholarship has seen the Hales judgement as an error, which unjustifiably appeared to sanction James's belief that he possessed a dispensing power.[39]

James's model of kingship also entailed the subjugation of the judiciary. Alfred Havighurst summed it up that 'to gain compliance in Westminster Hall he removed in less than four years twelve judges and appointed those who would serve his ends. The Bench was reduced to subserviency and aided and abetted the King in policies which the Bill of Rights subsequently denounced'. In January and February 1686, James dismissed two leading judges who opposed his use of the dispensing power, Baron Gregory and Judge Levinz – the latter was to exact his revenge during the trial of the bishops. Also dismissed was Heneage Finch, the solicitor-general, who not only opposed the dispensing power but also refused to obey James's order to draw up warrants confirming Catholic army officers in their posts.[40]

Following Compton's dismissal as dean of the chapel royal, James only communicated with him through Sancroft. This enabled the Archbishop to experience James's prejudices at first hand. For example, James told Sancroft to insist that French Protestant refugees in London diocese should conform to the liturgy of the Church of England and that Sancroft was to insist that Compton should do this. In the spring of 1686, James told Sancroft that he was, in future, going to appoint all naval chaplains through the Archbishop not, as was usual, through the Bishop of London. It was about this time that it was rumoured, possibly apocryphally, that James, in a rare conversation with Compton, had pointedly said he was more like a colonel than a bishop. Compton was said to have replied that he had indeed once drawn his sword in defence of the constitution and would so again if it became necessary.[41] Also in the spring of 1686, Danby, who had not travelled abroad for 30 years, left London for Holland. It seemed clear that he was discussing events with William of Orange.[42]

On 10 March 1685/6, James issued a general pardon in religious cases, it was cleverly timed so as to disrupt all church courts business against Catholics and prevent the punishment of those who had been found contumacious in the previous court sessions. As part of this, about 1,200 Quakers were released from gaol.[43] The tactic achieved its goal as in the next few months and years bishops reported a significant decline in ecclesiastical cases before their courts.[44] Worse still, James issued pardons to Dissenters convicted of attending conventicles. This led some Exeter Dissenters to claim that they had complete freedom from the Conventicle Act.[45] Sancroft's response was to enact a metropolitan visitation, indicating that he was not happy passively to accept the general pardon but was going to enforce discipline where dioceses were not well-managed.[46] On 25 March, James reissued directions to the archbishops 'to forbid Anglican clergy discussing the controverted points of doctrine in the pulpit', by which James meant sermons which attacked Catholics and Catholicism. John Evelyn recorded in his diary plenty of evidence of the numbers of such sermons.[47] However, Compton felt that the preaching of the clergy was the only defence against a rising tide of Catholic propaganda. When he next addressed his clergy, Compton told them that 'if we exalt the King's prerogative above the Law, we do as good tell the people ... [that] the King may ravish their wives, spoil their goods and cut their throats at pleasure'. He also made no bones of his fear that bad counsel would make the King believe that he possessed a dispensing power.[48]

This was not what James had expected following his order to the Archbishop. He clearly believed that Tory clergy, like William Sherlock, dean of St Paul's, and John Sharp, who was dean of Norwich and rector of St Giles in the Fields, London, had been specifically encouraged by Compton to preach anti-Catholic sermons. There was an element of truth in this, as Tillotson, Stillingfleet, Tenison, Patrick, Sherlock and Wake had 'formed an agreement' to preach such sermons with Compton's tacit support. James was also irritated by the anti-Catholic books that poured from the presses.⁴⁹ It seemed that Compton was deliberately violating James's injunction to the Archbishop not to engage in controversy in the pulpit. James certainly saw it in this light. In June 1686, James wrote indicating this to Compton and ordering him to suspend Sharp from preaching. Sharp petitioned the King to remit this order but James, feeling that he could not rely on the discipline of the bishops, was determined that Compton was to be punished. John Evelyn observed that 'all engines ... [are] now at work to bring in popery amain' and when he saw James's new Catholic chapel built in the palace at Whitehall he came away 'not believing I should ever have lived to see such things in the K. of England's palace'.⁵⁰

In April 1686, James had established an Ecclesiastical Commission to enquire into and punish ecclesiastical offences. The Commission had sweeping powers: it could summon individuals to appear and punish clergy with suspension, deprivation and excommunication. James claimed a precedent for the commission in his father's Court of High Commission, although some felt that a law of 1661 made such a body illegal.⁵¹ James planned that there would be three clerical commissioners: Sancroft, Sprat and Crewe together with four lay commissioners, Lord Chief Justice Herbert, Lord Chancellor Jeffreys, Lord President Sunderland and Lord Treasurer Rochester. Sancroft was appalled by the prospect of a state commission assuming his authority over the clergy. Sancroft pleaded with James to excuse him from membership of the Commission, owing to his age and ill-health. Sancroft used ill-health almost indiscriminately to avoid doing what he did not want to. But after a personal interview, it was clear that this was a feint and that the Archbishop would not serve on the Commission as a matter of principle, following which the King denied Sancroft access to the court.

An insight into the working of the Ecclesiastical Commission came from the Earl of Huntingdon's later statements about it. Huntingdon had succeeded Lord Rochester on the Commission in 1687, and, after 1688, was keen to defend his actions. Nevertheless, Huntingdon's defence was instructive. He claimed that 'this commission was represented to me noe way contrary to the Acts of Parliament'. Moreover, 'persons in those times lay under great difficulties after almost all the Judges of England had declined their opinion for the Dispensing Power, which was the root and foundation of these proceedings'.⁵²

The first work of the Commission, beginning in August 1686, was to act against Bishop Compton, for not imposing a punishment on John Sharp, for preaching anti-Catholic sermons. Indeed Sharp had preached 15 anti-Catholic sermons in May 1686 alone, most of which were pointed attacks on those Catholics in favour at court. When Sunderland had written to Compton asking why he had not acted against Sharp for ignoring the King's *Directions to Preachers*, the Bishop had ignored him. Ironically, while Compton was under this attack, Sharp readily agreed to desist from preaching to avoid further attacks on his diocesan. Compton made a grudging submission to Sunderland that he was investigating Sharp's preaching, but it was not enough.[53]

Compton, who appeared before the Commission four times, claimed he was already taking action against Sharp but had to allow due process, and permit Sharp to defend himself. Compton also claimed that only Archbishop Sancroft had the right to try and judge him for failure to carry out his diocesan duties.[54] Compton attended the meetings of the Commission with his brother, Sir Francis Compton, the Earl of Nottingham and a retinue of lawyers. Between them, they were able to testify how badly the Bishop had been treated. This was the first moment at which Nottingham revealed his disquiet with the King. He protested to James about the Ecclesiastical Commission; he also persuaded his father-in-law, Viscount Hatton, not to fill commissions for the place left vacant in Northamptonshire by James's ejection of his opponents.[55] But James had equipped the Ecclesiastical Commission with advice of his advocate-general, Dr Thomas Pinfold, who attended Compton's appeal against his sentence and remained for the private deliberations of the Commissioners to give them legal advice.[56]

Despite the political consequences, the Commission convicted Compton and suspended him from his episcopal authority for failure to act against Sharp.[57] In fact, the Commission was badly divided and it was only with heavy pressure to vote against their consciences that a majority was obtained to act against Compton. Rochester was only brought to heel by a combination of threats and flattery. Ironically, Sancroft's membership of the Commission would probably have affected the vote and negated the action against Compton.[58] Franklyn, the King's Proctor, resigned his office rather than carry the notification of his suspension to the Bishop.[59] In Compton's absence, Bishops Sprat of Rochester, White of Peterborough and Crewe of Durham were appointed commissioners to exercise episcopal jurisdiction in the diocese of London. They tried to soften the blow by ensuring that all clerical vacancies in the diocese were filled in accordance with Compton's wishes.[60] Paradoxically, when John Sharp appealed to Sunderland to ask the King to permit him to return to his parish from Norwich, where he had fled, he was given permission to do so. When he arrived back at St Giles, he found that the Catholic priests against whom he had preached had further undermined the Church, and his curates were openly challenged in their work. Consequently Sharp resumed his anti-Catholic attacks, though not from the pulpit.[61]

There is no doubt that the suspension of their bishop stiffened resistance to James among the London clergy, who were in contact with many clergy across England. Gilbert Burnet's agent in London reported that the London clergy were 'frightened mightily; for if they had another bishop they could no more carry on jointly the opposition they made to popery; by which they have raised a spirit against it all over the nation'. The shock of Compton's suspension was such that Symon Patrick was said to be willing to contemplate 'a remedy which at another time they would regard as a great evil'.[62]

The prosecution and suspension of Compton was also regarded with horror by Sancroft and the other bishops. There were rumours that Sancroft was likely to be the next target of the Commission. Certainly George D'Oyly claimed that Sancroft kept a paper by him to outline his legal grounds to challenge the Commission's authority, in case he needed it.[63] Princess Mary, for the first time interceded with her father, and said that she could not believe that Compton, her old chaplain and tutor, meant the King any harm. James's response to his daughter was a sharp rebuff.[64]

It seems likely that for Princess Anne also the suspension of her old tutor was a turning point. In the wake of it, she wrote to Mary that she would be firm to her faith 'whatever happens'. Given the unpopularity of Compton's suspension, the Ecclesiastical Commission felt so much on the defensive that it published *A Vindication of the Proceedings of His Majesty's Ecclesiastical Commissioners*. But Compton's stock as a Protestant martyr grew rather than diminished. In March 1686/7, Compton petitioned the King to lift his suspension, but the King's price was Compton's complete submission to him, to which the bishop would not agree. The Commission also became unpopular among the nobility when it investigated the marriages of the Duke of Norfolk and Earl of Coventry, both of whom were offended by its intrusion.[65]

Patronage was also a serious source of conflict between James and the Church. In July 1686, Sancroft, seeking to re-establish good relations, suggested to James some names of clergy who he regarded as suitable for preferment. James's response was to ignore Sancroft's suggestions and to advance Parker and Cartwright to the sees of Oxford and Chester. These two bishops were wholly unacceptable to Sancroft; indeed at one point Sancroft even considered refusing to consecrate them, Cartwright having knelt before one of the Catholic bishops sent to England.[66]

In December 1686, James ordered the Charterhouse, an almshouse on the site of a former Carthusian monastery, to admit a Roman Catholic, Andrew Popham, as a pensioner, in breach of the statutes, which admitted pensioners by election only. Sancroft delayed for more than six months before finally protesting to the King and refusing, as a Charterhouse governor, to cooperate in the election of Popham. Sancroft's opposition to James was also seen in his decision to licence Henry Wharton to preach throughout his province. This exceptionally wide privilege meant that Wharton, known for his opposition to Catholicism and his willingness to preach such views, could speak his mind from any pulpit in the province of Canterbury.[67]

Early in 1687, James finally dismissed his brothers-in-law, Rochester and Clarendon, and replaced the latter with Lord Tyrconnell, a Catholic Irish peer with little or no sense of judgement. To many, this seemed further to undermine the Church as Clarendon, in particular, was so strongly regarded as its advocate. It was also a move that brought Danby closer to open opposition to the King for, though Danby had no liking for Rochester, he viewed him as a bulwark against James's Catholic appointees. From that point on, Danby was in almost constant communication with William of Orange.[68] Though even by the start of 1688 he did not have a clear intention to depose James.[69] James's tearful comment, when he dismissed Rochester, was 'I cannot have a man at the head of my affairs who is not of my opinions'. The dismissal of Rochester and Clarendon stirred further anxiety among Anglicans, and the King seemed to grow unpopular. In February 1687, Bishop Cartwright had to admonish the inhabitants of Hulme in Lancashire for causing a riot when they shut their church on 6 February, which was the anniversary of James's accession.[70]

The dismissal of Rochester and Clarendon also marked the start of a new and more vigorous attempt to force through the repeal of the Test Act. Already James had dismissed scores of magistrates who he felt that he could not rely on to apply tolerant policies to Catholics. About 500 magistrates were replaced by Catholics.[71] But in some counties there were insufficient JPs; in Shropshire, for example the spring quarter sessions in 1688 were postponed for lack of JPs.[72] In December 1687, James began attempts to pick off individual politicians, in a campaign called 'closeting'. This was a series of interviews in which James closely quizzed leading laymen and ministers about converting to Catholicism in his closet. Sometimes it included pressure to vote for the repeal of the

Test Acts. Given James's singular lack of charm, the process was usually maladroit and often counterproductive. Rather than bringing James a stream of converts, it offended some of the ministers and officials so grievously that they were lost to James entirely. Admiral Arthur Herbert, for example, who had served under James and been promoted by him, was affronted by the pressure James brought on him to convert and chose to resign his commission as admiral. Lord Shrewsbury resigned his command of a cavalry regiment after a 'closeting' with the King.

In February 1687, William of Orange despatched Everard van Weede van Dijkvelt to London to estimate the strength of feeling against James. Van Dijkvelt was also charged with reassuring those with whom William had established contact that he was committed to the Church of England and to the Test Act but wanted a measure of toleration for the Dissenters. A number of people, including James's brothers-in-law, Clarendon and Rochester, spoke to van Dijkvelt of their support for William. Also among those who met van Dijkvelt were three leading Tories: Nottingham, Danby and Shrewsbury. Nottingham, always cautious and anxious not to rock the boat, sent van Dijkvelt a letter confirming that William was 'the person on whom they found their hopes'. Later in the year, Nottingham wrote to William expressing doubt that James's policies would come to anything, so great was the opposition.[73] In contrast, Danby promised his service to the Prince. Shrewsbury promised himself faithful to the Prince and Princess Anne and sent John Churchill to talk to van Dijkvelt. He pledged his loyalty to his religion and told van Dijkvelt he set his loyalty to the King at naught. He also hinted that he would protect Anne from her father if necessary. Like Danby, Churchill was in regular contact with William from this point on. Of course, as James's nephew and the consort of his heir, there need have been no treasonable intent behind such expressions of support.

At Easter 1687, rumours circulated that James had become an affiliate of the Society of Jesus. Certainly he had acquired a new confessor, Father Warner, also a Jesuit and rector of the Jesuit College at St Omer. But, even before this, James had twice delayed elections to Parliament, largely on the basis of intelligence that suggested he still would not obtain one which would cooperate with the repeal of the Test Act. In some desperation, James began a campaign of bribery, a strategy he had previously rejected.[74]

Few were intimidated by the King's determination. Edward Stillingfleet preached before James, on 10 April 1687, on a biblical text recalling Moses's repudiation of his adoptive grandfather, Pharaoh, preferring to endure the afflictions of his own people. Even Bishop Thomas Cartwright was able to see the message that it was better to adhere to the Church of England and endure affliction rather than support James.[75] The government's desire that there should be no expansion of the conflict between Catholics and Protestants was also expressed in the suppression of William Wake's *The Present State of the Controversie between the Church of England and the Church of Rome, or an Account of the Books written on both sides*. This book and its successor published a year later, *A Continuation of the Present State of the Controversy*, were denied entry in the catalogue of books at Oxford by government decree. This had the reverse effect, as Wake then found a publisher for the books which were more widely circulated than if they had been entered in the University catalogue.[76]

The attack on the universities

In 1687, at James's behest, the Ecclesiastical Commission turned its attentions to the universities. It was an attempt to break the Anglican monopoly and to open clerical education to James's Catholicising policies. In April 1687, the Commission summoned the vice-chancellor of Cambridge for refusing to admit Alban Francis, a Benedictine monk, to the degree of MA at the King's request. When the vice-chancellor mildly replied that the statutes required that an incepting master should take the required oath, which Francis would not, and that he could not act other than in conformity to the statutes, the Commission deprived him of office. Lord Chancellor Jeffreys directly quizzed the vice-chancellor, but was embarrassed by him; when Jeffreys asked whether Cambridge had ever previously refused a royal mandate, the vice-chancellor was able to answer that it had done so when Charles II had made a similar request to advance a Mr Tatwell to the degree of MA, Tatwell had also refused to take the oaths so the University had refused him the degree.[77]

In Oxford there were already two Catholic heads of houses: James had established a crypto-Catholic, John Massey, as dean of Christ Church and the master of University College, Obadiah Walker, had converted to Catholicism.[78] James had also issued dispensations for some Catholics to matriculate and graduate without swearing the usual oaths of conformity and subscribing to the Thirty-Nine Articles. In January 1687, Robert Charnock, a Catholic, was admitted to a fellowship of Magdalen College and, in March, Sidney Sussex College, Cambridge, was ordered to admit Joshua Bassett, another Catholic, as a fellow.[79]

Soon after, the fraught case of Magdalen College, Oxford began. The King had ordered the fellows to appoint Anthony Farmer as president and the fellows had petitioned the King not to ask this of them, since Farmer was a Catholic. James had made a bad choice in Farmer as it quickly emerged that he was a troublesome man and a drunkard, who had procured women for undergraduates and even that he had said he was only masquerading as a Catholic to obtain James's patronage.

James dropped Farmer and chose instead Samuel Parker, the Bishop of Oxford, who was widely regarded as a crypto-Catholic, and told the fellows to elect him. Instead, they elected John Hough as president. At the end of May, the Ecclesiastical Commissioners cited the fellows to appear before them. The Commissioners proclaimed Hough's election void and appointed Mr Wiggins, the Bishop of Oxford's chaplain, to the presidency. The Commissioners also demanded that the fellows sign the most obnoxious and grovelling apology to the King, all 26 fellows refused and were proclaimed guilty of disobedience to the King.[80]

The Magdalen College affair was remarkable for the alienation it effected between the King and perhaps the most loyal of all Anglican bishops, Peter Mews. Mews was an ardent royalist: he had been a cavalry commander at Naseby, where he was injured and captured, he was later a secret agent, during which time he evaded sentence of death. At the Restoration, he was made president of St John's College, Oxford, in 1672 he was consecrated bishop of Bath and Wells – where he excommunicated Dissenters to deny them the vote so as to guarantee Tory electoral victories during the Exclusion crisis – and in 1684 was elevated to Winchester.[81] When Monmouth invaded in 1685, Mews took charge of the King's artillery during the battle of Sedgemoor. Of all Anglicans and bishops, therefore, Mews could claim the credentials of the most loyal. As bishop of Winchester, Mews was *ex officio* visitor of Magdalen College, Oxford. When, in April 1687, the King nominated Anthony Farmer as president of Magdalen, the fellows sought Mews's advice. He replied that the fellows should proceed according to the statutes. On 15 April, when the fellows elected Hough, Mews cooperated with the College to spirit Hough to Farnham Castle within 24 hours, where he immediately admitted him as president. Mews told Hough and the fellows that he admired their determination. From this period on, Mews was committed to the opposition to James. Later in October 1688, when James reversed his policy and restored the president and fellows to Magdalen, Mews took some satisfaction in personally restoring Hough and the ejected fellows, leading a service of thanksgiving in the chapel and, in the Buttery Book, striking out the names of the intruded Catholic fellows.[82] The Magdalen College affair was, as G. V. Bennett noted, the test case of passive obedience. The fellows had asserted their lawful rights and passively incurred deprivation for refusing to act illegally. What good had passive obedience achieved, other than to demonstrate the illegality of the King's behaviour, and his indifference to passivity? 'It is difficult to overestimate the shock and anger which these events caused.'[83]

The first declaration

In April 1687, two months after he had issued a similar declaration in Scotland, James issued his first Declaration of Indulgence for Liberty of Conscience – dispensing with the Test Act and, in effect, abolishing the Church of England's monopoly over public office. James, somewhat disingenuously, claimed that the Declaration was issued under his dispensing power, which he held had been declared legal by the Hales judgement. The Declaration reflected James's disappointment with the Tory Anglicans and the bishops. In the same month that James issued the Declaration, he spoke to two Anglican bishops he trusted, Bishops Sprat of Rochester and White of Peterborough, of his unwillingness to suffer toleration. As a consequence, few believed that James was a genuine convert to toleration.[84] On 20 April 1687, Bishop Thomas Cartwright of Chester, a supporter of James and one who undertook secret assignments for the King, recorded '[the Bishops of Rochester and Peterborough] said they could not but remember how vehemently the King had declared against toleration and said he would never by any counsel be tempted to suffer it'.[85]

Initially, the first Declaration seemed to have an effect. John Evelyn noted that the Dissenters' meeting houses were packed and Anglican churches thinly attended.[86] But many of the published denunciations of the Declaration had an impact on men like Richard Baxter, who resisted the blandishments of the King. Those Dissenters who were wary of James were those of 'the highest moral authority'.[87] Burnet's agent in London reported that the bishops could rely on the support of the Dissenters.[88] James's objectives in the Declaration were guessed by the broadsheet balladeers, not least the author of the *Dialogue*, who wrote a pretended conversation between James and Queen Mary of Modena. In it, James was made to say

> A gaol delivery now must be
> To make all consciences be free

Not out of zeal, but pure design
To make Dissenters with us join
To pull down Tests and Penal Laws
The Bulwarks of the hereticks cause
The sly dissenters laugh awhile
They see where lurks the serpents guile
And rather than with us comply
Will on our enemies rely.[89]

James asked ministers to orchestrate addresses of thanks for the Declaration, and these were included in the *London Gazette*. In Norwich, Lord Yarmouth bullied the corporation into making its address, and in Carlisle dragoons were quartered in the town until an address was forthcoming. The trouble was that, in time, James appeared to believe his own propaganda and genuinely believed that toleration was popular with Dissenters, and that he could therefore dispense with the support of the Church.[90] But the Dissenters did not wholeheartedly welcome toleration from James. Roger Morrice believed that Dissenters would never agree to the toleration of Catholicism; he wrote that Dissenters preferred to 'remain under the persecution of the penal laws in the hopes and expectation of receiving at the proper period some alleviation, than by separating from the English church, to go surely the one after the other to the ground'.[91] Morrice probably exaggerated the Dissenters' unanimity on this point. Thomas Jolly, a nonconformist minister in Chester diocese, responded equivocally to the Declaration and was probably representative of many others. He wrote in his notebook that the Declaration was 'according to our desires and above our expectations' but he doubted its durability: 'considering some men's designs in it time will discover, god will defeat'.[92]

In one case, the propaganda campaign to thank the King for his Declaration backfired badly. In April 1687, the Bishop of Oxford, Samuel Parker, a strong supporter of James, arranged for the clergy of his diocese to subscribe to an address thanking the King for his Declaration. The address that was advanced by the clergy was a subtly subversive document. It noted that the King had promised to protect and maintain the Church and the bishops. The reasons given in favour of subscription to the address were that 'it may continue the King's Favour, whereas the omission [of an address] may irritate the Treasury to call upon the Fifth Bond for first fruits at full worth'. The second was that it would testify to their submission to the bishop, who 'required' their address. But the address included four reasons against subscribing, these were that the Declaration ought to be issued by Parliament; second, it 'herds' the Church 'among the various sects under the Toleration'; third, it ought to have had the endorsement of the Synod or Convocation of the Church and alone a diocesan address might suggest a division or schism, and finally, 'it forfeits the present reputation we have with the nobility, gentry and commonality of our communion which may tempt them ... to disgust us for our rash compliance'. The final words were that 'this address is no instance of canonical obedience' and, in an attack on Parker, that 'till Bishops at their confirmation declare what faith they are of' it was unclear whether the clergy were to be obedient to the bishop or maintain the historic communion of the Church.[93]

Though not published in the *London Gazette*, the Oxford clergy address was printed and widely circulated. James pressed on and demanded that the leading Dissenters, Henry Care, Vincent Alsop, Thomas Rosewell and Stephen Lobb, who were inclined towards him, should also drum up addresses of thanks. In the end, 78 addresses of thanks were sent by Dissenters. Anglicans were also urged to write such addresses. Sunderland told the bishops of Durham, Rochester, Peterborough, Oxford and Chester that they should arrange addresses from their dioceses. It was uphill work to persuade clergy to sign, and in the end only the dioceses of Lichfield and Coventry, Lincoln and St Davids produced addresses.[94]

While Halifax has been censured for his timidity during the Revolution, he can be credited with some foresight. His *Letter to a Dissenter*, was written after the Declaration and published secretly in August 1687 in huge quantities – 20,000 poured from the press in six editions. In it, Halifax argued that the Dissenters' duty was to guard against Catholicism. He urged them not to be taken in by James, and wrote of the Declaration: 'this is a violent change and it will be fit for you to pause upon it, before you believe it.'[95] He also predicted, that when James failed to get what he wanted from the Tory Anglicans, he would turn to them for support. Halifax also persuaded the Dissenters that the Church was no longer committed to persecution of them.[96]

Occasionally, James made concessions to the Church of England, though rarely other than at the behest of an ally. In July 1687, prompted by an appeal from Dr Hooke, vicar of Leeds, and forwarded by Bishop Cartwright, the King conceded that the Declaration did not legitimise the non-payment of tithes and church rates, and ordered that suits against nonconformists for the recovery of tithes could still proceed.[97] In the summer of 1687, the London clergy, stimulated by the Declaration, were in the forefront of opposition to the King. Bishop Ken wrote in July 1687:

> For my owne part ye Thoughts of our Clergy, ye London Clergy especially, doe daily Rejoice me, I give God thanks for them, and I pray for them, They are pious and learned and zealous, all yt wee could wish them to be God's Holy Name be praisd for it, Certainley God who has sent so many Labourers into His field has designed ye Harvest to be plentiful, yt though they may saw in teares, yet they shall reape in joy.[98]

William Wake even suspected that some London clergy were involved in the planning for William of Orange's invasion.

In the autumn of 1687, James began a royal progress to persuade the country that he was no danger to them. At the same time, James felt he could offer prayers at some of the notable Catholic shrines in England and Wales, including St Winefriede's Well in Flintshire. The tour included Bath, where he touched for the 'King's Evil', and proceeded through the Marches to Chester, before returning to London via Oxford. As early as June 1686, it was rumoured that James had dismissed his domestic chaplains at Windsor and had conducted a 'ceremony of healing' for the 'King's Evil' with Catholic priests in attendance.[99] James was later to touch for the 'King's Evil' at Winchester, where 250 people were touched.[100] Servants were sent ahead of the royal progress to ensure a warm welcome. But the truth of it was that James was disappointed at the reaction to both his Declaration and his tour. Besides plans to restore Catholics to supremacy in Ireland, he had achieved little. No wonder Bonrepaus, the French ambassador, believed James was completely out of touch with feeling in the country.[101] In September 1687, during his tour of the country, James visited Oxford. He had a bad-tempered meeting with the fellows of Magdalen at which he said they had not behaved like gentlemen. He also showed his hand by saying 'You have done very uncivilly to me...you have affronted me, know I am your King and I will be obeyed. Is this your Church of England loyalty?' He then shouted, 'Go back and show yourselves good members of the Church of England...Get you gone, I command you to be gone, go and admit the Bishop of Oxford'. When the fellows again refused, James left Oxford in fury.

Three of the Ecclesiastical Commissioners visited the College in October 1687, to rebuke the fellows for ignoring the Anglican doctrine of passive obedience. Finally, the Commission expelled all the fellows from office. Lord Huntingdon, one of the Ecclesiastical Commissioners, later claimed that 'it cannot be denied the delegated Commissioners had much more to answer for in their proceedings than we who were at that time no commissioners for visiting Magdalen College'.[102] The Privy Council – with an eye to Sancroft's licensing of Wharton – also added the penalty that the deprived fellows were to be incapacitated from receipt of any ecclesiastical office. Trelawny ignored this by presenting one of the fellows, Charles Penyston, to a living in his gift. Five other fellows were rewarded in this way.[103] The punishment of incapacitation seemed to some to be a serious invasion of the fellows' property rights.[104] Some bishops allowed the ejected fellows the right to appoint to College livings. The bishop of Gloucester accepted the nomination to a Magdalen living in his diocese from the ejected fellows, and then reported the living no longer vacant to the intruded Catholic president and fellows.[105] Only Lord Chief Justice Herbert protested that Hough's election as president had been regular and therefore could not be contested. For his role in the affair, as an Ecclesiastical Commissioner, Bishop Crewe was nicknamed 'the grand inquisitor'. There were even rumours that Crewe had become a Roman Catholic, certainly he attended Catholic ceremonies.[106] In fact, Crewe was keen to suppress the Catholic chapels in his diocese and later sought the King's aid to do so.[107] At Oxford, the installation of the Catholic president of Magdalen was attended by only two fellows and the porter threw the keys at him, and no blacksmith in the city would agree to force the locks of the president's lodgings. It is clear that James's attitude to the universities was taken to be 'stark proof of James's ill-will to the Church' since it was designed to challenge the Anglican monopoly.[108]

When James arrived back in London, Dr Jane, the Professor of Divinity at Oxford, preached a sermon at Whitehall on 'the integrity and patience of Christians in difficult times'. Jane argued that God's providence enabled Christians to bear the most wicked tyrants. The only response was to live a holy life and to persevere in religion, and Anglicanism, in particular, since it was 'the most pure, primitive and true of all professions of Christianity under heaven'. This was exactly the sort of reluctant but compliant Anglicanism that reassured James: it advanced the claims of the Church of England, but at least it had the merit of not advocating resistance to the King.[109]

The attack on the landed alliance

In the face of stubborn resistance to him, James realised that the only hope he had was to manage the election of a Parliament in which he could force through the repeal of the Test Acts. But electing a sufficiently pliant Parliament was a tall order, especially given the dusty response he had received in 1686, when he last made the attempt. He had already tried in the summer of 1687 to get the assize judges on their circuits to sound out members of parliament on the issue of repeal of the Test Act and 'to feel their puls (sic) on this matter'.[110] But the response was not positive. James's solution was to institute 'the three questions'. In October 1687, lords lieutenant were instructed to put three questions to all parliamentary candidates and magistrates – they were fiercely enjoined not to delegate these duties to deputies. First, if they were appointed to office or elected to Parliament would they favour the 'taking off' of penal laws including the Test Act? Second, would they use their influence to elect candidates committed to this policy? Third, would they support James's Declaration for Liberty of Conscience by living amicably with Catholics? The strategy was another misjudgement and the lords lieutenant of Buckinghamshire, Worcestershire, Staffordshire and the North Riding of Yorkshire refused to cooperate in asking the questions. In Warwickshire, the lord lieutenant, Lord Northampton, told the magistrates that he would not himself answer positively to the questions. In Ripon, the magistrates refused to answer the three questions and in Leeds the same evasive form of words was used by all the JPs, suggesting that there had been some collusion in the response to the questions.[111]

Not only were lords lieutenant unwilling to put the three questions, but where the questions were put the results were predominantly negative.[112] It was said that in Norfolk and Wales only about a dozen candidates altogether were asked the questions. In the East Riding of Yorkshire, 21 were asked the questions, and 19 replied in the negative. There was, of course, no penalty for replying in the negative, but a commission of regulation under Sunderland was formed to scrutinise the answers of those who were members of corporations so that they could be removed from office, and therefore excluded from the franchise, by *quo warranto* amendment of borough charters.[113] In the first year of his reign, James issued 16 charters to Cornish boroughs alone.[114] Lord Bath told the King that however many times he amended the charters he would not get a different answer. So candidates for parliament and the magistracy freely acknowledged that they did not accept any of the three propositions. An analysis of magistrates who were asked the three questions indicated that only 16 per cent of Protestant JPs agreed to them, the rest were negative or absented themselves or answered ambiguously or gave mixed answers. London livery company members were also asked the questions, and 3,500 expelled from the companies as a result of their failure to agree.[115] Indeed at the Lord Mayor's show in 1687, the crowd remarked that all the 'jolly genteel' liverymen had been replaced by Catholic 'fanatics'.[116] The ambassador of the Holy Roman Emperor observed that the tactic 'has done more harm than one can express, seeing chiefly that nearly everywhere a negative reply was given'. Sunderland told the Papal Nuncio that, even if James did get a compliant Parliament, there was no guarantee that it would be willing to betray the Church.[117]

To compound matters, James then alienated his natural Tory allies by dismissing the lords lieutenant who had refused to cooperate with putting the questions. These included the Dukes of Somerset and Norfolk, the Earls of Shrewsbury, Derby, Pembroke, Rutland, Bridgwater, Thanet, Northampton, Scarsdale, Abingdon, Gainsborough and Bath. Many of these nobles were men whose fathers and grandfathers had suffered grievously for James's father. In February 1687, Sir John Reresby reported that every post 'brought news of gentlemen laying down their appointments and Papists, for the most part, being put in their places'.[118] Trampling on these noblemen gave James no advantage. In time, James would regret making enemies of such natural supporters, many of them quickly turned to William both before and after his landing. In a piece of petty spite in December 1687, James removed Bishop Compton's nephew, Lord Northampton, from the lieutenancy of Warwickshire for his failure to enforce the repeal of laws against Catholics.[119] In their place, Catholics, or crypto-Catholics were appointed as lords lieutenant, including Father Petre, Lord Dover and Lord Chancellor Jeffreys.

As the three questions would not secure James a compliant Parliament, James continued the process of restricting the urban franchise to those who would vote for his candidates, by amending borough charters. A wholesale revision of charters to narrow the vote to those James felt he could rely on was a complex and detailed business. Up and down the land, Dissenters in particular were let into the franchise. It was a strategy that had the potential to backfire as badly as the three questions. Some towns simply would not be cowed. Reading is a good example: James revised the charter only to find that the new corporation was no more inclined to his will than the last, so the charter was revised again. Between March and September 1688 alone, 35 warrants for new borough charters were issued.[120] But Lord Bradford told Sunderland that even if James could pack the Commons, the Lords would not agree to the repeal of the Test Act. To the discomfort of all, Sunderland joked that Lord Churchill's regiment could be called to the Lords to make sure of their votes – it was not clear if he meant that the troops would fire on the peers or be created peers to vote through the measure.

Besides pressing Catholicism on the universities, and his closeting campaign, James made no bones of his desire that the nobility should consider conversion to the Roman Church. To the surprise of those who knew of their earlier flexible religious views during the Interregnum, the Duke of Norfolk, the Earl of Shrewsbury and Lord Lumley refused to convert. But James was determined to have Catholic ministers. In the place of Lords Clarendon and Rochester he appointed the Catholics Lords Balasyse and Arundel, who were 72 and 78 respectively. When the Earl of Mulgrave was made Lord Chamberlain, it was also thought that he would convert, since he was known to have modest religious views. But Mulgrave resolutely said that he found it hard to believe that God had made man, but to believe man had chosen God was impossible. In fact, Mulgrave was an example of a politician sympathetic to James, who worked closely with the King, but saw his weaknesses. By 1688, Mulgrave 'assisted openly all the Protestant clergy'.[121] Rochester had been thought likely to convert, but he did not. James's Catholic advisers were sources of scurrilous gossip. To the disgust of some nobles, there were rumours that some Catholics at court doubted the legality of the marriage of James to Anne Hyde, the mother of Princesses Mary and Anne. The French ambassador was certainly persuaded by the evidence, and Lord Rochester was forced to find the officiating parson and obtain his signed testimony of the legality of the marriage.[122]

Meanwhile the advance of Catholicism in London went on apace. Catholic schools were founded, by the Jesuits in the Savoy, the Franciscans in Lincolns Inn and a Benedictine house was established in Clerkenwell.[123] The Papal Nuncio was received at court, and there were rumours of the restoration of a Catholic hierarchy. Corporations of all sorts, legal, municipal and educational were pressurised to admit Catholics.[124] At the same time, the army was housed at Blackheath, and was regarded by some as a threat to those who might oppose such developments.[125] In fact James had attempted to make the army more Catholic by the introduction of more Catholic junior officers in 1686 and 1687, but failed on both occasions, the army coming near to mutiny.[126] They had the example of the Irish army, in which 67 per cent of soldiers were Catholic.[127] None of this might have appeared quite so threatening if the Catholics had not been so keen to convert Anglicans. Thomas Tenison had been engaged in a wrangle in his parish, St Martin in the Fields, with a Catholic priest, Andrew Pulton, who was recruiting parishioners. Eventually, in 1687,

Pulton challenged Tenison to a debate, or 'conference' which, though it was held in private, attracted widespread attention. The conference gave way to a bad-tempered pamphlet war – including Tenison's own 83 page account of the meeting, which was published with the Archbishop's imprimatur. It was not long before the issues raised in the tracts were being addressed from the Jesuit pulpit in the Savoy. During the debate Luther's principles were called 'filthy', Tenison was attacked as a 'bigot' and aspersions were cast on Pulton's education and spelling. Conferences like this were not uncommon between Anglican and Catholics and Anglican and Dissenters. Tenison was able to draw some satisfaction from attracting an Irish Capuchin priest, John Taffe, to the Church of England, in June 1688, during the climax of the seven bishops' trial.[128] He also gained a reputation for his generous sermon, in November 1687, on the death of Nell Gwyn, Charles II's 'Protestant whore' who, Tenison claimed, died a penitent.[129]

Tenison's own views on Catholicism were well-known since the publication in 1683 of his *Argument for Union*. In this he wrote

> The Romanists are a mighty body of men...they are all united in one common polity, and grafted into the one stock of Papal headship...they have varieties of learning, they pretend to great Antiquity, to Miracles, to Martyrs...to extraordinary charity...they have the nerves of worldly power, that is, banks of money and a large revenue. They have a scheme of policy always in readiness.[130]

Such views damned Tenison in James's eyes, but he had a wide following in London. Moreover, James's policies had convinced Tenison that both Church and country needed to resist the King. For this reason, Sancroft invited Tenison to attend meetings with the bishops on the issue of the Declaration. The bishops had only to look to Scotland, where James was pushing toleration of Catholics much harder than he was in England, to see what might happen.[131]

3
The Confrontation

The gathering storm

During August 1687, the King continued to issue dispensations to Dissenters freeing them from penal laws. By the end of the month it was assumed that any Dissenter who wished to do so could receive a dispensation against the Clarendon code, including the Test Act. The number of applications was so great that an office was established to deal with them; the usual rate was 50 shillings for an indemnity for a whole family. This, it was assumed, would drive sufficient wedge between the Church and Dissenters.[1] In November, James replaced the Anglican mayor and aldermen of London with Dissenters and urged them to hold their own services at the Guildhall chapel. But to the King's disgust they attended regular Anglican services at the chapel, communicated and took the oaths required by the Test Act, and permitted 5 November to be celebrated in high Protestant style with the usual anti-Catholic revels. In Gloucester, James tried to prevent the city from lighting bonfires to celebrate the gunpowder plot and the mayor restricted the bonfires to two; in defiance of the King, the bishop permitted two more in the cathedral precincts.[2]

By October 1687, Henry Compton, who had retired to Fulham during his prolonged suspension from his diocese, felt that he had reached the point of no return. He wrote to William of Orange in guarded but unmistakable terms:

> The terms by which you were pleased to express yourself in reference to the Church of England were every way so obliging and satisfactory that I look upon myself as bound in duty to acknowledge the deep sense I and every true member of the same Church ought to have of

so great a blessing. And though you are at present at a distance from us, and not so well able to partake of the fruits of so good intentions, yet when we shall have served this King with all fidelity, so long as it shall please God to continue him amongst us; as none that you know will question the sincerity of your performance, so I make no doubt, but you will soon find the benefit of having taken up so wise resolutions. For, Sir, you that see all the great motions of the world; and can so well judge of them, know there is no reliance upon anything that is not steady in principles, and profess not the common good before private interest. I pray God to continue to be gracious to you, and to direct and prosper all your counsels, and to crown the endeavours of your life with the consummation of all happiness.[3]

Compton did not know at the time that James was considering calling the Convocation of the Church to demand that it formally acknowledge the legality of his dispensing power.[4] In time the King thought better of the idea.

In November 1687, came the news that Queen Mary Beatrice, James's wife, was pregnant. Not everyone was convinced by it, even the French ambassador seemed sceptical, writing to Louis XIV that she 'believes herself pregnant'. Previously, the Queen had a series of miscarriages, but everyone was aware of the implications of the birth of a male heir. Such an heir would displace Princesses Mary and Anne and William of Orange in the line of succession, and would inevitably be raised as a Catholic. What worried many was the determination of James's supporters that this child would be a boy. When she heard the news of the pregnancy, Princess Anne was said to be speechless with rage, and unable to disguise her fury. James used the opportunity to make a final attempt to persuade Mary to understand his reasons for conversion. She sent an evasive response. The Queen's pregnancy also created a quandary for some of the moderate Catholic peers, Belasyse, Powys and Dover, who knew that, if the Queen miscarried again, Mary would remain the heir presumptive. This was a difficult balancing act. For others, the Queen's pregnancy was a tipping point; Sunderland told the King that he had decided to convert to Catholicism but that he would not do so immediately, preferring a time when the King might gain some benefit from such a high profile defection to Rome.[5] In such circumstances, it is difficult not to regard Sunderland's later conversion as a cynical move.

In the same month, James established a commission under the former Dissenter and Catholic convert Sir Nicholas Butler, a London merchant, to supervise election business and, in particular, the further revision of borough charters and the identification of trustworthy candidates for seats. Butler's agents were equipped with horses and paid 20 shillings a day to find local people who could act as correspondents for court candidates. The role of the agents was also to find people who would disperse royal propaganda and report intelligence to the court. Agents frequently toured the boroughs, being more likely than county seats to swing to James. G. H. Jones argued that the electoral agents were probably more successful than previously thought by historians. Their reports may have been optimistic, but even of county seats it was estimated that 223 out of 312 might elect candidates loyal or sympathetic to James.[6] Douglas Lacey argued that the electoral strategy in the autumn of 1687 and spring of 1688 was sufficiently successful for moderate Dissenters, and even Anglicans, to respond to the electoral tactics and return to James's side.[7]

By Christmas 1687, the contacts between William and disaffected English politicians were becoming the stuff of rumour; James ordered postal services to Holland to be searched. Consequently, Gilbert Burnet arranged for his nephew, James Johnstone, to establish a network of accommodation addresses in England, and to use secret codes, invisible inks and revealing solutions. These were the instruments of a plot. But it was clear that William needed some way of communicating with Englishmen on a wider scale. In mid-January 1688 therefore, England was flooded with copies of a pamphlet called *Their Highness the Prince and Princess of Orange's Opinion about a General Liberty of Conscience*. The pamphlet contained a letter from Grand Pensionary Fagel of Holland to a Scottish refugee from James's policies. Fagel's letter was drafted by William and its title clearly indicated that it contained William's views. These were that William and Mary favoured freedom of religion for Protestant and Catholic Dissenters but within an establishment in which the Test Act gave Anglicans a monopoly on public office. Despite desperate attempts by James to exclude the pamphlet – including searching coaches and wagons leaving London – it gained wide circulation throughout Britain. Richard Andrews, James's election agent in the West Country, reported 'Mons Fagell's and other pamphlets are spread through all parts to prejudice those inclined to your Maj'ie'.[8]

The *Opinion* revolutionised the scene in London. Dissenters who might have been inclined to support James turned to William; he offered the prospect of sincere toleration rather than toleration as a mere expediency to liberate Catholics, and one that might only therefore be temporary. The Earl of Devonshire wrote to William that the people were in 'raptures' at the tract. A furious James wrote to the States-General and demanded that the English regiments, stationed in Holland for their defence, be returned to England. At the same time, crude attempts were made to try to force the moderate Tories to return to James. Danby, Halifax, Shrewsbury, Dorset and Henry Sidney all received anonymous notes warning them to 'make your peace with the King, or be assured that after this 27 of January you have not many days to live'.[9]

In January 1688, James asked the Pope to divide England into four districts, each with an episcopal vicar-apostolic. This was the return of the Catholic hierarchy that so many feared. James promised an income of £1,000 for the new bishops. James's choice as president of Magdalen, Bishop Samuel Parker of Oxford, died in March 1688 and the King chose Bonaventure Giffard, one of the new vicars-apostolic, to replace him. Even Catholics, including one of the other vicars-apostolic, John Leyburn, believed that this was a step too far. But James pressed on, and it was widely assumed that Magdalen was being prepared to be a new Catholic seminary to train priests to convert England. Worse was the fear that since James had been bold enough to appoint a Catholic to the bishopric of Oxford, the same would happen to the vacant diocese of York. But James's Catholic allies were far from united. The moderate Catholics, Lord Powis and Lady Middleton, were fiercely opposed to the spread of Jesuit influence; some Catholics accused Sunderland of being in secret contact with William; Petre and Sunderland were rivals for influence over the King and Petre believed Sunderland had blocked his appointment as a cardinal, something for which James had been consistently lobbying the Pope. On Sunderland's advice regarding the lack of evidence of a chance of obtaining a compliant Parliament, James postponed the elections until May 1688.

In February 1688, responses from the county seats to the three questions were returned to London. It was clear that there was much less support in the counties than in the boroughs. Moreover, in some boroughs the numbers of voters to be removed were so great as to be a serious problem. In Westminster it emerged that the franchise would have to be slashed from 2,000 to 200 to produce a compliant electorate.[10] In an attempt to get submissive boroughs, the charters of Northampton, Leicester and Leominster were reissued four times and that of Maldon six times. The management of the charters soon became chaotic, so much so that Winchester, which had had its charter withdrawn in 1684, was regulated as if it was still in force – and the law officers of the crown were unaware of the changes imposed on the city.[11] This was, in part, a reflection of the casual way in which James treated the boroughs. When the King visited Chester, Bishop Cartwright told him of some aldermen who opposed the Declaration; James nonchalantly replied 'let me know what alderman opposed and I will turn him out'.[12] Moreover, in some revised charters, such as those issued to Barnstaple, Bridport, Chester and Wells, the entire corporation was removed at a stroke.[13]

The issuing of new charters to boroughs exposed James's position: he clearly had to amend charters in such a way as to indicate which groups he sought to place in control of boroughs, such as Catholics and Dissenters. There could be no claim to an unprejudiced motive for revision. Second, with each new charter came 'a Clause of Pardon and Dispensation' which freed members from the legal requirement to take the oaths of supremacy and allegiance. Third, the charters usually invested the right to elect MPs only in the corporation and swept away other franchises. Finally, the charters also reserved powers to the King to remove individual members of each corporation.[14] These features revealed more than many others the lengths to which James was prepared to go in order to obtain a Parliament he wanted and to displace Anglicanism as the established church.

In February 1688, James also issued a proclamation for suppressing seditious and unlicensed books and pamphlets. He ordered that only those who had been licensed by the King or a bishop, and had served an apprenticeship, could print and sell books. It was, claimed James, because a series of treasonable books and pamphlets had 'disturbed the minds of our loving subjects'.[15] One book, which the King licensed, sought to demonstrate the schismatic nature of the Church of England. It concluded: 'the Protestants are guilty both of material and formal schism; since 'tis evident they have done both a schismatical fact and out of schismatical affection'.[16]

James felt obliged to follow-up the disobedience of lords lieutenant in the issue of the three questions. Already some had been sacked, but in February 1688 a further round of dismissals occurred: Lord Dorset was removed as lord lieutenant of Sussex and Lord Oxford dismissed from Essex. To save the King the bother of dismissing him, the Duke of Newcastle resigned from the lieutenancies of Nottinghamshire and Northumberland. By early March 1688, the management of packing the new Parliament had become such a mess that James was forced to postpone elections again until the autumn.[17] James was also aware that something was going on between William and dissidents: when Danby's son asked for permission to travel abroad, James told him he could go to the continent but not travel to Holland.[18]

James's attitude to his dispensing power had been presaged in February 1688, when he reissued his Declaration of Indulgence in Scotland, which, he said, was issued by dint of 'our absolute power ... which all our subjects are bound to obey without reserve'.[19] Two months later, the second Declaration was issued in England. Within a few days Gilbert Burnet anonymously published *Reflections* which included the comments:

> The King's suspending of laws strikes at the root of this whole government and subverts it quite ... [though] the executive power of the Law is entirely in the King; and the Law ... has ... made it unlawful upon any pretence whatsoever to resist it ... [yet] the Legislative power is not so entirely in the King. No Law can be either made, repealed, or, which is all one, suspended, but by the consent [of Parliament].[20]

By Easter 1688, Mary had conceded 'there is no other way to save the Church and the State than that my husband should do to dethrone [my father] by force'. And Danby told William that 'our zeal for the Protestant religion does increase every day in all parts of the nation and the examination of the minds of the nobility and gentry had made such a union for the defence of it throughout the Kingdom'.[21]

The second declaration

When, in April 1688, James reissued his Declaration for Liberty of Conscience, he added the requirement that the Declaration should be read in all parish churches in the country. The Declaration was, by his own admission, drafted by Bishop Cartwright,[22] though there were also suggestions that Father Petre wrote it.[23] Later there were rumours, said to originate with Father Orleans, one of James's Jesuit priests, that it was the Presbyterians who had first suggested the idea of making the bishops enforce the reading of the Declaration by the clergy in the churches of their dioceses, as a means of splitting Anglicans and Dissenters. It seems unlikely that most Presbyterians would have suggested this, but it is possible that it came from those who were close to James.[24] The renewal of the Declaration was a reflection of James's failure to obtain the election of a Parliament which would pass the repeal of the Test Acts.

As Halifax had predicted, since the Church of England had been a grievous disappointment to James he turned to Dissenters. Roger Thomas concluded James's purpose in the Declaration and the order to read it was 'to form a coalition of Roman Catholics and Dissenters that would be powerful enough either to cow the Church of England into submission or break it if it refused'.[25] This seemed to be confirmed by the first statement of the newly appointed Catholic bishops who, in a pastoral letter to the Catholic laity, claimed: 'a great part of the Nation, whose persuasion in points of religion doth differ most from yours, and which in time past hath been severe upon your persons is willing to enter into a friendly correspondence with you'.[26] James was also astute enough to recognise that his Declaration had the potential to stall the growing contacts between Anglicans and Dissenters and ruin talks on measures to bring them closer together. James may have made promises to protect the Church of England but few now placed any weight on them. James also pursued other policies designed to woo Protestant Dissenters to his side. He advocated progressive economic policies including coinage reform, the prevention of the export of wool and outlawing imprisonment for debt, which would attract those in trade and commerce, as Dissenters often were.

In Catholic circles there was some unease. James's Catholic ministers, Lords Powis, Arundel and Balasys, had warned him that toleration was a mistake.[27] And in Catholic areas there was some reluctance to challenge the Church of England. In Preston, for example, the Tories and Catholics were loyal to the King, but permitted the performance of *The Devil and the Pope*, an anti-Catholic comedy. In Liverpool, the Catholic dominated corporation was unwilling to support the moves to repeal the Test Act.[28]

An important element in the failure of the Dissenters to offer strong and unalloyed support to James was the profusion of theological works which emphasised the apocalyptic danger of Roman Catholicism. Besides the works of William Wake and Thomas Tenison, in 1687 and 1688 the works of James Ussher were reprinted, which warned that Catholics would trample on Protestantism and subjugate it in England.[29] These works achieved considerable purchase on Dissenters. Perhaps the most significant work was the *Cases of Conscience*, published in 1685 by the leading London-Anglican clergy to persuade Dissenters to return to the Church.[30] The *Collection of Cases* addressed the anxieties of Dissenters, not just in presenting the Church of England as strongly differentiated from the Roman Catholic Church, but by showing how much Dissent and Anglicanism had in common doctrinally and liturgically. Written by, among others, William Sherlock, Thomas Tenison, John Sharp, William Cave, Edward Fowler, William Claggett and Symon Patrick, the *Collection of Cases* had a dramatic effect in reassuring Dissenters that Anglicans, who had previously persecuted them, were now their Protestant brethren, and emphasising the importance of Protestant solidarity.[31]

Even before James's confrontation with the bishops, it is clear that William's mind was moving towards intervention. At the end of April 1688, William was anxious about James's attitude towards Holland and the Low Countries, and wrote to Arthur Herbert that if he was invited by 'men of the best interest' to rescue the Protestant religion, he would be ready to do so in September.[32] Among the London clergy and episcopate there was increasing evidence of anticipation of intervention by William. William Wake wrote of William Clagett 'I have reason to think Dr Clagett knew something of it' since in the winter of 1687 he had ominously discussed with Wake 'putting a stop' to the King's policies. Beddard believed that it was impossible that the Anglican hierarchy did not know of the correspondence between England and The Hague. Bishops Compton of London, Lloyd of St Asaph and Trelawny of Bristol were all 'secret favourers of a foreign interest'.[33] When rumours of the Prince of Orange's invasion spread, William Wake concluded 'I soon perceived that our Great Divines generally favour'd this Cause'.[34]

James's Order in Council, issued on 4 May, required bishops to send the reissued Declaration to their clergy, with orders for it to be read in all London churches, and those within a 10 mile radius of the city, on two Sundays, the 20 and 27 May, and in country churches – those beyond a 10 mile radius of London – on 3 and 10 June. James may have felt to be on strong ground in commanding its reading, his brother had ordered the reading of declarations during his reign, though he had not required the bishops to enforce them.[35] James's grandfather had insisted on the reading of the *Book of Sports* in the 1620s, but it was not an auspicious precedent, having led to widespread clerical disobedience. However, the Declaration of 1688 communicated James's claim to a royal prerogative to dispense with the penal laws and Act of Uniformity, leaving every man free to worship God according to his own conscience. It was both constitutionally and religiously controversial. The Tory Charles Hornby described it as 'the utmost stretch of obedience'.[36] Michael Mullett wrote that the enforcement of the reading of the Declaration 'seemed to require the Anglican clergy to act as accomplices in the assassination of their own church'.[37] James's suggestion that the Declaration be read in Dissenting meetings also was abandoned on the advice of ministers who felt that it would further unite Dissenters and Anglicans.

The Declaration coincided with the fact that James had sent an envoy to the Pope to explore whether Catholics could be admitted to English dioceses. It also coincided with the fact that the four Catholic vicars-apostolic (Bishops John Leyburn, Bonaventure Giffard, Phillip Ellis and James Smith) made it clear that their goal was the conversion of the King's Protestant subjects to Catholicism.[38] Moreover, after Leyburn had been on a tour of the north, he claimed to have made 21,000 converts, all of whom he had confirmed.[39] James ensured that he captured the initiative: his supporters on the bench of bishops, Cartwright of Chester, Crewe of Durham and Sprat of Rochester were called together to issue welcoming addresses for the King's second Declaration. Cartwright wrongly advised James that most bishops would permit their chancellors and archdeacons to take the decision of whether to read the Declaration.[40] Late in April, Sunderland made a tactical error in sending a draft address to Trelawny, who sat on it for three weeks until it was too late to issue. Bishop Trelawny had the foresight to discuss with leading clergy of his diocese his intention not to sign the address, he noted that 'this was necessary for several, as 'twas hinted to me, out of feare would otherwise have signed it altho' now they refused it to a man'.[41]

Despite their half-hearted response to the first Declaration, James also asked some leading Dissenters to present an address of thanks to him for the renewed Declaration. These Dissenters included William Penn, Sir John Baber, the King's physician, Alderman Daniel Williams, Samuel Slater, Richard Mayo, Stephen Lobb and Vincent Alsopp. They gathered at the house of John Howe on 23 May, to discuss whether to send the King an address of thanks. The King, hoping to pressure them into sending him such thanks, sent word to the meeting that he was waiting for their thanks in his closet and would not retire until he heard from them. Daniel Williams was instrumental in preventing Dissenters from signing the declaration of thanks. Williams recounted to the Dissenters his experiences in Dublin under Catholic persecution, and he told his fellow Dissenters that he was far more afraid of Popish persecution than royal threats. Fearing that the objective of the Declaration was Catholic persecution, the meeting refused to send the King their thanks.[42] The Papal Nuncio reported that the attempt to divide the Anglicans and Dissenters had failed and that the whole Church 'espouses the cause of the bishops'.[43] It seems clear that the subsequent petition of the seven bishops reflected the stiffening resolve of the Dissenters' not to thank James for the Indulgence.[44]

The confusion in which the Tory Anglican churchmen found themselves was reflected in a poem published on 28 May 1688 which ran

> You topping clergy of the English Church
> Who still would leave the Dissenters in the lurch
> And by your laws always have them perplext
> And with your pinfolds have them sorely vext;
> Whom you have persecuted unto death
> And so will do (we fear) while you have breath...
> Passive obedience now is gone astray
> Prayers and tears which was the only way...
> This is the very temper of all such
> Who are true Tories in your English Church
> But by refusing of your free consent
> What do you mean, or what is your intent
> In this affair, except it be to show
> You will be cross, unless your Church shall crow?[45]

It was also reflected in the vulgar comment by Lady Hervey, speaking to some bishops: 'you have made a turd pie, seasoned it with passive obedience and now you must eat it yourselves'.[46] Another poem 'The Clerical Cabal' noted that

> Tho of passive obedience we talk like the best
> 'tis prudence, when interest sways, to resist.[47]

The importance of the Declaration, and the Order in Council requiring it to be read, was that they brought the nature of James's rule into every corner of the country. In every parish the parson, in being required to read the Declaration, had to ask himself whether he accepted James's dispensing power. And, if he did accept the dispensing power, would he read a Declaration which destroyed the privilege of his church? In the order to read the Declaration, James made the issue of his kingship no longer one that only affected politicians and those in the metropolis, it affected every parson and his parishioners. The King's Declaration also affected every parish with a conventicle or meeting house. In part this was why the bishops decided that they had to provide leadership; without the protection of the bishops' petition, the clergy would have been alone and forced to wrestle with the issue of supporting their King's prerogative claims or their Church. This was widely perceived as the explanation for what happened. On 26 May 1688, John Gadbury wrote to Sir Robert Owen in Wales that the bishops' petition was 'to excuse the contempt' of the clergy in refusing to read the Declaration.[48] Lord Lonsdale said that the only argument in favour of reading the Declaration was to obey the King, the argument against it was that it would destroy the establishment of the Church of England.[49]

The Declaration was not sent to the bishops, on the grounds that they might claim not to have received it, but was promulgated in the *London Gazette*, denying them such an excuse. The politicians, Rochester, Halifax and Nottingham all refused to speak against the Declaration. In fact the pusillanimity of the lay politicians is a remarkable feature of James's reign before the trial of the seven bishops. Only Clarendon advised the clergy not to read the Declaration; Rochester advised compliance; Halifax 'was so very cautious that he would give no advice at all', and Nottingham wavered first advising against defiance and then only to defy the King if there was complete unanimity.[50]

Canvassing opposition

The response of the leading churchmen to the Declaration demonstrated their capacity for political strategy and tactics. Within a day of the issuing of the Declaration, leading London clergy of all hues, Tillotson, Patrick, Stillingfleet, Tenison, Sherlock and Fowler, were active in talks with both Anglican clergy and Dissenters. Fowler, Tillotson and others, who had been contributors to the *Collection of Cases*, were especially influential among the Dissenters.[51] As part of the London consultation, a paper entitled 'A Comprehensive Sense of the Clergy' was written and widely circulated in the city. This made clear that there was no 'want of tenderness' to Dissenters behind the Anglican resistance to the Declaration of Indulgence.[52] Most Dissenters had already seen through James's ploy to use the Declaration to split the Dissenters from their Protestant brethren in the Church. Roger Morrice was one of those Dissenters who also saw the issue of the dispensing power as central to the defiance of the King. Morrice held that to comply with the dispensing power was 'most Heynously criminall to publish the Prince's private will and pleasure against his Legall and incontrovertible will'.[53]

The leading London clergy met in a series of secret meetings on 4, 7, 8 and 11 May. It was reported that some clergy were 'frequently in their clubbing and caballing, as if ye reign of Shaftesbury were designing to be among us'.[54] The initial meeting of London clergy seemed to support submission to the King's will, until a message from a leading

Dissenter was read. In Oxford, Humphrey Prideaux commented of the Declaration, 'things look cloudy'.[55] Certainly there is some evidence that opinion was mercurial, even in London. On 13 May there was a meeting at which more were for reading the Declaration than were against it. It was at this meeting that the tide turned, however, largely because of the intervention of Edward Fowler, vicar of St Giles Cripplegate and a friend of the Dissenter Roger Morrice.[56] Fowler, hearing the opinion of the clergy, said

> I must be plain. There has been argument enough. More will only heat us. Let every man say Yea or Nay. I shall be sorry to give occasion to schism, but I cannot in conscience read the Declaration; for that reading would be an exhortation to my people to obey commands which I deem unlawful.

The leading London clergy, Stillingfleet, Patrick, Tillotson and Sherlock agreed with Fowler and led the others.[57] Simon Patrick recalled that, after two further meetings with London clergy, they were all agreed that they should ask the bishops to petition the King against reading the Declaration. Patrick, Tenison and 20 others travelled the length of the city and concluded that 70 clergy of London, a significant majority, would agree not to read the Declaration.[58] Simon Patrick wrote down the names of those London clergy who would not read the Declaration and gave it to Bishop White to give to Sancroft.

The results of the canvass were reported to Sancroft on 17 May.[59] Sancroft was clearly relieved that the Dissenters had not been fooled by James's apparent conversion to toleration. It was James's great achievement that he was able to bring Sancroft to friendly feelings towards Dissenters.[60] Sancroft had also dined with Lord Clarendon and the Bishops of London, Ely, Peterborough, Chester and St Davids. The last two – known supporters of James – 'discomforted' the conversation during the meal, but withdrew soon afterwards, leaving the others to discuss the Declaration. It was astute to wait until they had left, since Bishop Cartwright was clearly advising James and doing so in concert with the Bishop of St David's. Once alone with fellow bishops and friends, Sancroft indicated that he could not promulgate the Declaration. The following day was spent in consultations with other bishops. On 18 May, the seven bishops met Tenison, Stillingfleet, Patrick, Tillotson, Sherlock and Grove. Sancroft had called a meeting of bishops at Lambeth, his riders were sent to all but the most cringing supporters of James. He told the bishops not to tell anyone 'that you are sent for'.[61] They prayed and discussed the Declaration. Simon Patrick indicated that the clergy believed that the bishops should lead the resistance, although Lowther said that the London clergy were agreed that if the bishops would not resist James they would do so alone. Perhaps therefore there was a sense that the bishops felt obliged to lead. Hornby said that the bishops believed that 'a general calamity' was likely if the clergy resisted James without the leadership and protection of the bishops, because the clergy were more vulnerable than the bishops. Thus it was, in part, that the pressure from the London clergy impelled the bishops to the decision to lead opinion rather than follow it. Nevertheless Sancroft tested this opinion by asking for solid support among the London clergy.[62]

Petitioning the King

Sancroft took the view that it was not part of the duty of the clergy to promulgate the Declaration, and as James was not a member of the Church of England, he viewed his motives with suspicion. He had previously been urged to defy the Declaration by Clarendon, though Maurice Ashley's claim that it was the Earl who 'screwed up the Primate to action' is unlikely.[63] After dinner on the evening of 18 May, the petition was drafted and written out by Sancroft and signed by him and Bishops Lloyd of St. Asaph, Turner of Ely, Lake of Chichester, Ken of Bath and Wells, White of Peterborough and Trelawny of Bristol. Compton, though in full agreement, felt his suspension meant that he could not sign the petition; Mews was ill, and Frampton arrived too late. Later signatories were Frampton on 21 May, Compton and Lloyd of Norwich on 23 May, Ward on 25 May and Lamplugh on 29 May.[64] But as a patristic scholar, Sancroft was quite happy with the idea of seven bishops, seven being a sacred number in the Bible. After a long and detailed discussion, the bishops resolved to address a petition to the King. The full text of the petition was

> The humble petition of William, archbishop of Canterbury and of diverse suffragan bishops of that Province now present with him, in behalf of themselves and others of their absent brethren, and of the clergy of their respective dioceses,[65]
>
> Humbly sheweth,
>
> That the great averseness they find in themselves to the distributing and publishing in all their churches your Majesty's late declaration for liberty of conscience proceedeth neither from any want of duty

and obedience to your Majesty, our Holy Mother, the Church of England, being both in her principles and constant practice unquestionably loyal nor yet from want or due tenderness to Dissenters, in relation to whom they are willing to come to such a temper as shall be thought fit when that matter shall be considered and settled in Parliament and Convocation, but among many other considerations from this especially, because that declaration is founded upon such a dispensing power as hath often been declared illegal in parliament, and particularly in the years 1662, 1672, and in the beginning of your Majesty's reign, and is a matter of so great moment and consequence to the whole nation, both in Church and State, that your petitioners cannot in prudence, honour or conscience so far make themselves parties to it as the distribution of it all over the nations, and the solemn publication of it once and again even in God's house and in the time of His divine service, must amount to in common and reasonable construction.

Your petitioners therefore most humbly and earnestly beseech your Majesty that you will be graciously pleased not to insist upon their distributing and reading your Majesty's said declaration.[66]

It was the nub of the bishops' view that James was laying claim to more than a particular dispensing power, he wanted to assert the general right to suspend laws. This was the difference between the particular dispensing power allowed on specific occasions by the *Godden v Hales* judgement, and a universal right to dispense or suspend laws at will. The bishops turned aside from more cautious counsel, particularly that of Nottingham, who advised that the worst thing would be for some clergy to read the Declaration and some not to, indeed Nottingham believed that it would better for all the clergy to read the Declaration than a split in the Church be created. One of William's supporters said that Nottingham's caution 'almost ruined the business'.[67]

Although the events of May 1688 have been presented by historians as focused entirely on the metropolis, it is clear that all across the country men and women knew the dilemma facing the bishops and clergy and were anxious to see what would happen. John Romsey, Sir Charles Kemys's servant, wrote from Bath on the day the petition was presented, telling Kemys that Bishops Trelawny of Bristol and Ken of Bath and Wells were going to London – they had already arrived of course – and that opinion was divided as to whether they intended to confront the King or debate the matter with him. Romsey reported that in Bristol the view was that 'their business was uncertain'. Romsey also reported the scare story that 25 Dutch men of war had been seen off the coast and that it was thought that Princess Mary had appointed a representative to attend the Queen's confinement. Both of these was false.[68] Mary Woodforde, the wife of Dr Samuel Woodforde Canon of Chichester, a staunch Anglican, who had been appalled by the Monmouth Rebellion against James, wrote in her diary on the day of the trial of the bishops:

> This day is a day of trial for our good bishops, they being brought up before the Council for not reading the King's Declaration for Liberty of Conscience contrary to their judgement. Now Lord make good thy promise to them, and give them a mouth which all their, and our, enemies may not be able to gainsay nor resist. And be forced to say of a truth the Lord is with us, and in us, to which end dear Lord give us all true repentance for all our sins.[69]

The interest the population took in the issues confronting the bishops was also reflected in cheap and popular broadsheets that poured from the press. One of these, *The Learned and Loyal Abraham Cowley's Definition of a Tyrant*, quoted Cowley's view that 'I call him a tyrant, who...having a just title to the government of a people, abuses it to the destruction or tormenting of them'. It went on to pose questions such as 'whether the legislative power be in the King only, as in his Political Capacity, or in the King, Lords and Commons'? Could the King make a pronouncement against a law of Parliament?[70] One London correspondent, sending news to Wales, copied out long passages from another broadsheet, *An Apologie for Such as do not read ye Declaration...*, which circulated in London by the end of May.[71] In this way, the issue of compliance or resistance to James spread across the country as a result of the bishops' petition and trial.

When they had drawn up their petition after dinner at Lambeth on 18 May, despite the lateness of the hour, the six prelates passed over the Thames to Whitehall without Archbishop Sancroft, who again claimed to be in ill-health and, in any case, had been refused access to the court. The aim of the six bishops was to hold a preliminary conference with Lord Sunderland, to tell him of their intention to petition the King and to ask that he excuse them from reading the Declaration. They hoped to get a later appointment to present the petition, and asked Sunderland to read the petition himself, so that he could explain its contents to the King beforehand, to avoid taking

James by surprise. If Sunderland had done so, he might have softened matters so as to avert a collision between the King and the bishops. But Sunderland, probably aware of the significance of the deputation, refused to look at the petition, and arranged for the King to see the bishops straight away. Thus it was that the petition was presented late at night by Lloyd, Bishop of St. Asaph, instead of Archbishop Sancroft.

When they were introduced into the royal closet, the six bishops knelt and presented the petition. The King received it and, looking at it, said,

> 'I know this hand, this is my lord of Canterbury's handwriting.'
> 'Yes, sir, it is his own hand,' replied the bishops.

The King then went on: 'This is the standard of Sheba.[72] What? The Church of England against my Dispensing power! The Church of England, They that always preached it!' The King realising their intention to resist his order, folded it up, and said, 'This is a great surprise to me; here are strange words. I did not expect this from you. This is a standard of rebellion'. He repeated this phrase more than once during the interview. According to one account, James said 'I have heard this before, but could not believe it. You look like trumpeters of rebellion; you aim at my prerogative. But I will not lose one branch of it. Take your course and I will take mine, my commands will be obeyed, do it at your peril.'[73]

When the King mentioned rebellion, Lloyd replied: 'we would lose the last drop of our blood, rather than lift up a finger against your Majesty', and this sentiment was echoed by the others.

'I tell you this is a standard of rebellion, I never saw such an address', repeated the King.

Trelawny fell on his knees, exclaiming, 'Rebellion, sire! I beseech your Majesty do not say so hard a thing of us, for God's sake! Do not believe we are or can be guilty of rebellion! It is impossible for me or any of my family to be guilty of rebellion! Your Majesty cannot but remember that you sent me down into Cornwall to quell the Monmouth rebellion, and I am as ready to do what I can to quell another, if there were occasion.'

Bishop Lake added: 'Sir, we have quelled one rebellion and will not raise another.'

Turner went on: 'We, rebel, sir? We are ready to die at your feet.'
Ken added, 'Sir, I hope you will give the liberty to us, which you allow to all mankind', meaning the right to petition the monarch.
White said: 'Sir, you allow liberty of conscience to all mankind; the reading this Declaration is against our conscience', which seemed a way of hoisting James by his own petard.
The King replied: 'I will keep this paper. It is the strangest address which I ever saw; it tends to rebellion. Do you question my dispensing power?', and then, more pointedly, he said, 'some of you here have printed and preached for it, when it was for your purpose.'
'Sir,' replied White 'what we say of the dispensing power refers only to what was declared in Parliament.'
The King replied 'The dispensing power was never questioned by the men of the Church of England.'
Lloyd boldly replied 'It was declared against in the first Parliament called by his late Majesty, and by that which was called by your Majesty.'
James again repeated that he regarded the petition as a rebellion, and said he insisted on having his Declaration read by the clergy. He tried to argue that there was a difference between the right to suspend acts of parliament and to dispense with them in certain cases. This was a fine distinction, and one which he had previously ignored.
'We are bound', said Bishop Ken 'to fear God and honour the King. We desire to do both. We will honour you; we must fear God.'
James, with increasing fury, asked, 'Is this what I have deserved, who have supported the Church of England, and will support it? I will remember you that have signed this paper. I will keep this paper; I will not part with it. I did not expect this from you, especially from some of you. I will be obeyed in publishing my Declaration.'
'God's will be done' said Ken.
'What's that?' demanded James.
'God's will be done' said Ken and White murmured his agreement.
The King dismissed them in anger, with the comment: 'If I think fit to alter my mind I will send to you. God hath given me this dispensing power, and I will maintain it. I tell you, there are seven thousand men, and of the Church of England too, that have not bowed the knee to Baal'. This final being a reference to his 7,000 strong army.[74]
Whatever the feelings of James and the bishops that evening, in the words of John Dryden, 'the passive Church had struck the foremost

blow'.[75] On reflection, James felt that the bishops had played into his hand and that this was an opportunity to drive a wedge between Anglicanism and Protestant Dissent.[76] Later accounts blamed James's Catholic advisers for the fury of the King's response to the bishops: 'the non-compliance of the Bishops boiled in the stomachs of Father Peters and the rest of the Gang.'[77] It might have boiled more furiously if they had known that William and Mary had written their congratulations to Sancroft on the stand he had taken.[78]

The leak of the petition

What made the confrontation much more damaging was that later that night what was claimed to be copies of the bishops' petition, protesting against the dispensing power, were offered for sale by hawkers through the streets of London, a practice which was then without precedent. It was said that the printer who produced it made a £1,000 from his night's work.[79] Meanwhile, the bishops returned to Lambeth by the archbishop's barge. The King, who was exceedingly offended at the publicity given to what he had regarded as a private conversation, assumed it was a further outrage on the part of the prelates. He sent a stern rebuke to Sancroft, complaining of the publication as treason. Sancroft replied with an expression of regret and surprise at what had happened, and protesting

> ignorance of the matter, and great perplexity as to how the petition could have got abroad, since he had written it out with his own hand to prevent any treachery on the part of a secretary, so that there was no copy, only the original document, and that was in his Majesty's own possession.

It is clear that the petition must have been sent to press before or immediately after the bishops left the King. As the audience did not start till ten o'clock, and within two hours their petition was bawled about the streets, there were three people who might have circulated the petition: Lloyd, Compton or Sunderland, assuming Sancroft was not responsible for it.

Agnes Strickland felt it was probably the work of Sunderland, to whom James, though he pocketed the petition, would almost certainly have given it.[80] The problem is that unless Sunderland was deliberately making trouble for James, or foolishly believed the public would support the King, he had no motive to arrange for the printing and sale of the petition, and he tended to act as a restraint on James's more forceful instincts. It seems unlikely therefore to have been Sunderland who leaked the petition. Tindal Hart claimed that Lloyd did not have time to hand a copy of the petition to a printer.[81] However, there were a number of manuscript copies of the petition in the papers of both Bishop Lloyd and of William Williams, the solicitor-general. Lloyd certainly had a motive to leak the petition – given his support for William – and is assumed to be guilty by John Carswell.[82] One of the copies in Williams's paper was made by William Bridgeman, a clerk to the Privy Council, on the evening that the petition was presented.[83] Whilst it is unlikely that James leaked the petition – not even he was so incompetent as to believe that broadcasting the petition would earn him sympathy – it is possible that Bridgeman might have. Lloyd ambiguously hinted that he believed James was responsible. In a paper, written in Sancroft's hand, but intended to be spoken by Lloyd at their trial, Lloyd claimed that, since James had kept the petition, he 'might have supprest our Petition; and if he had but declar'd that such was his pleasure, there would have been no copie of it now remaining in the world'.[84] Was it that had James told the bishops not to copy and publish the petition, Lloyd would not have done so?

Edward Carpenter and David Horsford suggested Compton as the likely source of the printer's copy of the petition. It is a strong claim; Compton had been present at the meeting at Lambeth but did not sign the petition nor did he go with the bishops to see the King. So he was at liberty to take a copy to a printing house and get it distributed through London. Perhaps, with time on his hands and having already detached himself from James, he saw an opportunity to create trouble for the King.[85] Compton also had the requisite ruthlessness to do so. J. R. Jones took it as certain that Compton was to blame in an attempt to force a confrontation.[86] G. H. Jones has suggested an alternative view, which is that the copy of the petition circulated in print bears a resemblance to the 'Comprehensive Sense of the Clergy' circulated during the earlier consultation of the London clergy, and therefore the printing of the petition may have been the responsibility of a well-connected London parson.[87] Roger Thomas endorsed this view, suggesting that any London clergyman with access to the 'Comprehensive Sense' could have added the bishops' names to the bottom and sent it to a printer.[88]

Whoever was responsible for its publication, within a day or so Roger Morrice, for one, had a copy of the petition.[89] Already, despite their protestations otherwise, there is little doubt that at least two of the bishops, Lloyd and Trelawny, had already crossed the point of no return in their attitudes towards James. To Burnet's cousin, Trelawny was reported to have said 'If King James sends me to the Tower, I know the Prince of Orange will come and take me out, which two regiments and his authority would do'. Certainly Lloyd and Trelawny were in full contact with William by this time and had similar expectations.[90] Moreover, Lloyd, since the bishopric of St Asaph had no London palace, stayed with Lord Clarendon who was clearly alienated from James. Either way, the publication of the petition made the breach between the King and the bishops irreconcilable and placed them on a more determined collision course.

The propaganda war

Having lost the first blow of the propaganda war, James was quick to respond. Through Henry Hills, his official printer, James issued a tract, entitled *An Answer to a Paper importing a Petition of the Archbishop of Canterbury and Six other Bishops, to His Majesty, touching their not Distributing and Publishing The Late Declaration for Liberty of Conscience.* The *Answer* gave chapter and verse of the confrontation with the bishops, including a copy of the petition. It then worked through the grounds on which the King expected their obedience, citing precedents from Nathan, Zadok, Solomon and David to Bacon's *Essay on Subjection*. It railed at the episcopate for not emulating the bishops of Charles I's reign, accused them of failing to support liberty of conscience for their Dissenting brethren and defended the dispensing power of the monarch. In a moment of prescience, the *Answer* commented: 'trust is the sinew of society... and distrust the disbanding of it'. It concluded that no man should set up conscience against duty and smash the two together.[91] But this was bravado; James vacillated on the issue of whether he ought to bypass the bishops and sent his order to read the Declaration directly to the clergy. On a number of occasions printers were asked to print such instructions, but each time the order to the printer was recalled.[92]

James's propagandists were more effective in a tract published a few days later, 'with allowance', which put the King's case in a fully developed way. *The Examination of the Bishops Upon their Refusal of Reading His Majesty's Most Gracious Declaration...* claimed that the King's clemency in granting the Declaration made him 'the nearest pourtraict of that Deity whose vicegerent he is'. The King, it argued, had 'resolved to break the fetters that extort 'em, the penal laws'. The Church ought to abandon her 'sometimes darlings' and it should be an obligation on her to abolish the penal laws. After some historical meandering through the Reformation, the *Examination* sought to separate the issue of the faith of the king and the loyalty of the people, it asked: were Protestants traitors under Catholic kings? Why should Catholics be traitors under a Protestant regime? It also sought to align the Churches of England and Rome: the Anglican Church did not require reordination of Catholic clergy who became Protestants, so Catholic orders must be valid. Such arcane and ancient issues seemed to bedevil the times, 'what contradictions and cobweb laws are here' claimed the tract. But sweeping away cobwebs entailed asking 'what rubbish is here put together to build the great fence of a Church with' and claiming that the penal laws were 'hypocrisy and imposture' against Protestant Dissenters since it was only when the Dissenters grew in number that the Anglicans were persuaded that the 'Old Arts' of persecution should be revived to crush Nonconformists with 'imputations of oppression, cruelty, sedition and riots'. The Church, in implementing the Clarendon code, was 'undoubtedly guilty of more barbarity than the ten primitive heathen persecutions'.

The *Examination* also pondered on issues of fallibility and asked how a Church that rejected infallibility could have the assurance that its actions against Dissenters were reasonable. It argued that the duty on the Church of England to throw off penal laws was 'a greater duty' than even the King's or the Dissenters. It also argued that it was fruitless for the Anglicans to resist:

> For whilst the Government continues in the Hands of a Prince of the Romish Religion, those [penal] Statutes will utterly lie dead; for the Royal Indulgence, a Prerogative in the Crown, will never put them in execution: And if abolisht, however the next Protestant Prince has the power of Resumption, if his Conscience shall think fit to give them a Resurrection.[93]

The King was resolved to 'eternalise his Glory' and to be the 'Pater Patriae' of other kings by making 'the lion and the lamb lye down together'. The *Examination* could not resist arguing that Catholic judges were no more incapacitated in administering justice than Protestant. Nor could it forbear commenting that it was offensive to the King to find that his own household officers had to swear the Test, and therefore he was 'to be debarred the choice of its own menials'. It defended the

King's amendment of charters: why should not the King have charters 'to his own heart's liking', for this would not cause floods of Catholic voters to come forward.

The climax of the *Examination* was an attack on the bishops: 'our conscientious mute prelates were certainly under the falcination of no ordinary hot zeal, or something else as warm, to be warped into disobedience in so poor a cause'. And 'that the illegality of the King's dispensing power, should be a spectre that appeared so dreadful to their Lordships, yet walks invisible to every mortal eye-sight else, is not a little surprising'. Surely the King's dispensing power in the penal laws was merely the power of clemency? Surely the 'noli posequi' in criminal cases was just like the dispensing power? So, seemed to be the argument, what was all the fuss about?[94]

James's propagandists also joined in fighting the bishops with their own arguments. At the same time as the *Examination* appeared, another tract was issued entitled *Toleration Tolerated*.[95] It claimed to be extracted from the works of Bishop Jeremy Taylor and asserted that the different sects of Christianity were evidence of the 'great fault' of mankind. It also compared toleration as a private indulgence of people to hold different views with toleration as a public licence of a sect. But it did not shirk open propaganda for the King:

> The case stands thus, His Majesty having in a manner wrought a miracle to Appease his people, viz to make the winds blow from all corners at once, that all their respective vessels may sail together, and none may ever want a wind to launch out at his pleasure, one would think they should all be satisfied... The truth is there has been a disease a long time in Government, and now it comes to be canvassed into, and searched, the pain in curing makes some men love the Disease better than the Remedy.

The tract commended the King as sincere and wanting nothing but the good of his subjects. It also claimed that the case was a contest of consciences and that the Bishops wanted to restrict consciences whereas the King wanted to free them.[96]

A third tract, *An Address to His Grace the Lord Archbishop of Canterbury*, was also thinly veiled government propaganda.[97] Addressed to the bishops, the tract adopted the King's pained tone:

> you know best how unexpected your late address was to his Majesty, who had contributed so much to the placing of most of you in your

Episcopal Chairs, and would so willingly have believed you were of the same Loyal Temper you expressed in his Royal Brother's reign. How his Majesty resented it, I hope you will reflect upon, because you ought not to forget it.

With an eye to the Monmouth rising, the tract expressed surprise because 'you have heretofore in times of trial manifested not only your loyalty, but your Zeal for the person of the King'.[98] James was depicted as seeking 'a true and lasting brotherly affection and harmony' between his subjects of different religions, and with an eye to the Dissenters' financial interests it claimed such concord would 'by the encrease of trade enrich the whole nation'. The issue of toleration was conceded as a moot point, two parliaments having ruled against it. But the bishops' declaration against the King's dispensing power and disobedience to his command was unconscionable. It was claimed that James had combined declarations in defence of the Church with his Declaration of Liberty of Conscience.

The principal claim of the *Address to the Archbishop of Canterbury* was that toleration 'neither deprives your Lordships of your dignities, nor robs you or the inferior clergy of their possessions; nor doth it hinder the free exercise of your Religion'. This, above all, demonstrated that James did not understand the basis of the objection to the Declaration. There was also a veiled threat, the bishops were asked to consider 'whether this disobeying of the King be a likelier means to preserve your Lordships in your places and dignities and the Church of England in its lustre, than your obedience would have been?' It would antagonise the Dissenters, claimed the author. There was also a veiled accusation of cowardice, if a single man had been ordered to read the Declaration he would have done it, but 'nos numeri sumu' encouraged the bishops to defy the King. And worse, hinted at threats: 'Methinks I hear the sad and loud groans of Thousands of those who are desireous to preserve their religion, and the King's clemency to the Church'. so that all loyal men would tremble 'to think what fatal consequences this obstinate act may produce', especially when they gave some thought to the heavy fines Henry VIII laid on the clergy. The consequences for the bishops themselves also seemed terrible, they had

> raised a disobedient, if not a rebellious spirit of opposition against the King through all his dominions; you have inflamed the minds of the multitude to oppose the King's desires...your lordships have divided yourselves from a loyal, judicious and religious part

of the Church...you have divided yourself from the whole body of Dissenters.[99]

It was not possible for James to keep his propaganda away from a defence of Roman Catholicism. *Some Queries Humbly Offered to the Lord Archbishop of Canterbury and the Six Other Bishops concerning the English Reformation*, which was also officially printed at this time, was an extraordinary defence of Catholicism. Cranmer was claimed as a reformer who had been a Catholic, and it was asserted that 'Roman Catholic Bishops did also receive their Orders from Christ and his Apostles, and consequently are true bishops, and therefore to be heard'. Clearly this appeared to pave the way for Catholic bishops taking over Anglican dioceses. The *Queries* trod on dangerous ground – given the enormous popularity of Queen Elizabeth I – by asking 'whether Queen Elizabeth, born of Ann Bolen (Queen Katherine yet living) can be thought legitimate?' Supporters of the Reformation were dismissed as motivated by 'the Lucre of Church lands' and the King James Bible accused of such errors that England needed a new Bible.[100] James's search for, and appeal to, popular support through the use of government tracts and propaganda was, in the words of one historian, 'the point in the Revolution at which the appeal to the people was made, and James was stripped of moral authority'.[101] The Glorious Revolution, if presented as a coup d'etat, underestimates the commitment of James and the bishops to win popular support for their cause. James contributed to this contest, though he later expressed surprise when he lost it. Whatever the quantity of James's propaganda, it persuaded few.

4
The Tower

Silent pulpits

Two days after the bishops' confrontation with James was the date set for the reading of the Declaration in London churches. It was a bad error of judgement to permit the date for the reading of the Declaration in the rest of the country to be two weeks later than the date for the reading in London, not least because the country clergy were clearly looking for a lead from those in London. The London clergy were as good as their word. Simon Patrick claimed that it was not read in any parish church, and only in Westminster Abbey because Thomas Sprat, bishop of Rochester, a supporter of James, was dean and ordered a minor canon to read it – though the congregation walked out and Sprat absented himself in the country. At the Whitehall chapel, only a member of the choir was prevailed upon to read it, and afterward the preacher, the Latitudinarian John Scott, preached an archly apposite sermon on 'the vicissitudes of worldly things'. Evelyn claimed that the Declaration was 'almost universally forborne throughout all London'.[1] In Wales, it was reported that 'the King's edict for the reading the Declaration of Indulgence was wholly disobeyed'.[2] In fact, only four London parish clergy, one of whom was hopeful of appointment to the vacant see of Oxford, read the Declaration. Some leading clergy, such as Stillingfleet and Tenison, made sure they were out of London on 20 May, so that they could not read the Declaration.[3] In other cases, clever clergy read the Declaration to empty churches, having dismissed the congregations before doing so.

Word of the disobedience travelled far and swiftly. Edward Clark, the Whig exclusionist, received word from a friend that 'ye Declaration was read at four places in London at two of ye four ye people went out of ye church at ye time of reading of it', but he went on 'the King does intend to send the Declaration to every parson in England and ye refusers are to be turned out'.[4] The public interest in the reading of the Declaration was 'the tide of public excitement on which the Revolution rode'.[5] On 25 May, John Evelyn recorded that 'all the discourse now being about the Bishops refusing to read the injunction for the abolition of the Test Act...the action of the Bishops universally applauded and reconciling many adverse parties'.[6] The Declaration allowed every parson and every member of his congregation to participate in either opposition or submission to James.

Some of James's supporters were having doubts however. Even the Papal Nuncio told James that the clergy were in full support of the bishops, that there were no internal divisions between the Anglicans to be exploited, and their hopes of the Dissenters had 'vanished'.[7] Only in Somerset, where there was determined and stubborn Dissent looking for toleration, did some Dissenting ministers 'understanding that many of the Clergy of the Church of England would refuse to read the King's Declaration, thought it their duty (though not directly commanded) to read the same last Lord's Day in their Congregation'.[8]

The legal case

James was alert to the danger that some bishops and clergy would defy his Declaration. On 19 May, he had summoned the judges to Whitehall, and Clarendon claimed it was in response to anticipated defiance from the bishops. Lord Chief Justice Wright attended divine service at Serjeants' Inn on 20 May, and heard for himself the excuse from the clerk that he had forgotten to bring the Declaration with him to the chapel. In the wake of the disobedience of the London clergy, the issue of what to do with the seven bishops became much more significant. One of the copies made of the bishops' petition on the same night they delivered it to the King was sent to the solicitor-general, William Williams. Williams copied his thoughts directly onto the petition. Some of these were simply tendentious. Where, in the petition, the bishops referred to their aversion to publishing the King's Declaration 'in their churches', Williams wrote in the margin: 'their dioceses, they are not their churches'. Where the petition referred to the bishops not acting out of 'want of tenderness to Dissenters', Williams wrote in the margin: 'trickery insinuation of intended favour w. Dissenters'. Further down, mention of the dispensing power led Williams to observe 'ye Foundation of it is upon a dispensing power'. When the bishops claimed that they could not 'in prudence, honour or conscience' read the Declaration, Williams exclaimed 'against their prudence'. At the end of the petition, Williams noted that the bishops wanted the King to declare the Declaration illegal, he speculated whether they would consider relief for Dissenters in Convocation and Parliament. Williams also regarded the challenge to the dispensing power a 'reflection upon ye consequence of this upon ye prudence, honour and conscience of ye King'. He went on to list the precedents for libels, and for cases where 'a company of men' came to the King. Williams concluded that such a deputation 'doth wound the King'. What is clear from Williams's notes is that the solicitor-general was already putting into James's hands the legal means by which he could act against the bishops in the courts.[9]

It has been claimed that James's action against the bishops misfired because he had not sufficiently prepared the legal arguments he could advance.[10] In fact, it is clear that the legal case against the bishops had been meticulously considered in the weeks before the trial. Solicitor-General Williams evaluated the possible arguments that could be advanced against the bishops. Certainly Williams felt that the bishops 'did forme and contrive a plot...for the purpose of seditiously, contemptuously and unlawfully declar[ing] the same in the royall presence of the King'. Williams claimed to have evidence 'to prove their writings and contriving and writing of the paper...'; he could also prove the publishing of it in the King's presence and believed he could prove that they had further published it. He considered a prosecution, under a law of Richard II's reign, regarding the slander of the King. He also sought legal precedents for prosecuting them for condemning a royal proclamation. Williams argued that if the King's proclamation *had* been illegal they should have remained silent and then they might have been able to plead their case if they had not supported the reading of the Declaration. The most significant question Williams raised was 'if the king was right' how would he establish this outside a court of law? Williams clearly relied heavily on the Hales judgement and listed the ways in which that judgement could be used.[11]

On 1 June, James held a council meeting at which Sunderland spoke in favour of clemency for the bishops and of not pressing the confrontation to a trial. But all the Catholics on the council spoke in favour of the prosecution of the bishops.[12] James's biographer claimed that it was Lord Chancellor Jeffreys who persuaded the King that the petition was a seditious libel, though this is disputed by most studies of Judge Jeffreys.[13]

The empty country pulpits

On 3 June, the country clergy followed the lead of their London brothers. As one popular verse put it

> Canonical blackcoats like birds of a feather
> In town and country do flock all together.[14]

There had been a highly effective campaign to make the country clergy aware of their metropolitan brothers' opinions. William Sherlock was at the centre of this campaign and wrote its manifesto, entitled *A Letter from a Clergyman in the City, to his Friend in the Country*, dated 22 May 1688. The *Letter* was circulated in thousands, perhaps enough for every clergyman in the country. The *Letter* was also circulated in a very circumspect fashion, as it was clear that James was having posts searched to prevent exactly this sort of campaign. In Norwich, Humphrey Prideaux received 2,000 copies, enough for all parsons in the diocese. They had been sent from London to Yarmouth and then forwarded to give the impression that they had been printed abroad. The *Letter* argued that James's aim – through the evil counsel of his advisers – was to ruin the Church by opening doors to Catholicism. It also claimed that to read the Declaration was against the teachings of the Church; an honest parson could not read the Declaration therefore. Sherlock also claimed

> The Dissenters who are wise and considering, are sensible of the snare themselves, and tho' they desire Ease and Liberty, they are not willing to have it with such apparent hazard of the Church and State: I am sure that tho' we were never so desirous that they might have their Liberty (and when there is opportunity of shewing our inclinations without danger, they may find that we are not such Persecutors as we are represented) yet we cannot consent that they should have it this way, which they will find the dearest Liberty that was ever granted.[15]

Sherlock, seeking to create a community of interest, claimed that the nobility and gentry were on the side of the clergy 'who have already suffered in this cause', and argued that 'all good Protestants would despise and hate us' if the clergy read the Declaration. He denounced the Declaration for turning clergy into 'meer machines and tools to be managed wholly by the will of our superiours' and this contradicted the care the clergy had for the souls of men. In a postscript, Sherlock added mention of the small number of clergy in London who had read the Declaration.

The *Letter* was quickly contested by *The Countrey-Minister's Reflections on the City Minister's Letter to his Friend*. This belittled the claims that the nobility and gentry opposed the Declaration and that it would open the door to popery: 'I cannot imagine why our Church doors may not be shutt and lock'd as fast after reading as before'. Much was made of the claim that 'the Prince was set above the law...and the notification of his pleasure made anything...obliging'. There was also the significant question: 'is it not enough for them [the bishops] to desist from obeying it without making clamours against it in print?' It asked the clergy whether their refusal to read the Declaration meant that they had to agree with everything read in the churches? Did it also mean that the clergy claimed the right to 'pry into the King's secret intentions'. The *Countrey-Minister's Reflections* also contested whether the dispensing power and toleration were against the traditions and teachings of the Church.[16]

The letters for and against reading the Declaration continued to be published, even during the trial of the bishops. They gradually adopted a more aggressive tone. *The Minister's Reasons For His Not reading the King's Declaration, Friendly Debated by a Dissenter* was an obvious piece of James's propaganda. It referred to the bishops as 'those rigid prelates who made it a matter of conscience' and rehashed many of the arguments that had been previously advanced – including that the Declaration would let popery into the Church. It questioned why clergy were prepared to read letters sent to them by their bishops, but not those sent by the King. It also sought to duck the legal questions of whether the King or Parliament was right about the dispensing power, claiming they were 'both complicated and equivocal'. It asked, which of the Thirty-Nine Articles contradicted the Declaration. More pointedly, it reminded the clergy that Dissenters were frequently summoned to court and prosecuted for their consciences, yet the Anglican clergy used the claim of conscience to refuse toleration granted to them by the King. In fact, claimed the Dissenter, 'the Declaration is an incitement to all the King's subjects to worship God'. Above all, the Anglicans' jealousy towards other denominations had 'raised such a thick mist before your eyes, that no Royal Promise, no settled laws, no common interest can dispel it, so as to give a discerning of the proper means to arrive at a secure Establishment'. But the Dissenter concluded that the Declaration also included the recommendation to congregations 'the choice of such members to serve in Parliament, as may happily finish that which His Majesty has therein proposed and mercifully begun'.[17]

If the King's supporters appropriated the endorsement of the Dissenters, the bishops' supporters advanced the claims of French Protestants who had fled into Germany. Letters claiming to be from some French ministers, and translated from the French, 'pronounced in favour' of the bishops who had 'exceedingly well answered the duty of their charge' in supporting Protestantism. The French Protestants even-handedly accepted that the bishops had behaved harshly to Dissenters, and the Dissenters had been wrong to separate from the Church of England and argued that, in principle, the bishops' insistence on reordination of French Protestant ministers did not well dispose them to the bishops; but the refusal of the 'venerable prelates' to cooperate with James had 'highly justified themselves from the reproach that was laid upon them of being Popishly affected, and of persecuting the Dissenters only but of a secret hatred to the Reformation'. The bishops had become standard bearers of Protestantism:

> Could you see those faithful servants of God disobey the order of their sovereign, expose themselves thereby to his disgrace, suffer imprisonment, and prepare themselves to suffer anything rather than betray their Consciences and their Religion?[18]

The French ministers claimed that the King's aim was the establishment of Popery and the 'extinction of the Reformed Religion in England'. This was 'a very great evil, and such as all true Protestants are obliged with their utmost power to oppose'. Dissenters were urged to suppress their sectarian interests at such a time in support of the general Protestant cause. The strength of the French ministers' letter was the examples they could cite from their experience of 'the cruelties which it has so lately exercised against the Churches of Hungary, of France and of the Valleys of Piedmont'. Louis XIV had made exactly the same promises of freedom of conscience in the Edict of Nantes that James was making in the Declaration, and these were promises which the Pope freely allowed the Catholic rulers to dispense with. In short 'we ought not to believe ... King James II'.[19] Moreover the French Protestants' voices appeared to be independent, disinterested between the Church and Dissent, but firmly supporting the bishops.

Inevitably, there were diverse reactions to the order to read the Declaration across the country. Simon Patrick, in Peterborough, wrote 'I did what I was able to prevent the reading of it; which was not difficult to persuade them unto, they being generally everywhere inclined to follow the example of the London clergy'.[20] Bishop Crewe saw to it that the Declaration was read in much of Durham diocese, but when it was read in the cathedral, John Morton changed the words of the responses from 'Oh Lord save the King' to 'Oh Lord save the People'. In the diocese of Durham, 21 clergy read the Declaration, and 30 were suspended for failure to do so by Crewe; in Oxford diocese only 6 read the Declaration.[21] Though George Hickes estimated that in the dioceses of Oxford, Lichfield and Coventry and Hereford not more than four or five in each diocese read the Declaration.[22] In the diocese of Lincoln, Bishop Barlow sent it to his clergy, but in many places the gentry resisted the reading; at Horton, Leicestershire, Parson Theophilus Brooks found that the patron, Lord Ferrars, 'swore it would not be read in his Church'.[23] In the diocese of Worcester, Bishop Thomas did not distribute the Declaration on the grounds of conscience. He wrote to Sancroft:

> I have retained in my custody the pacquet of the printed coppys of the Royal Declaration of Indulgence, which I could not transmit to the clergy of my diocese ... It is a piercing wounding affliction to me to incur his Majesty's displeasure ... and dread of the indignation of the King of Kings.

Bishop Beaw of Llandaff wrote to Sancroft, on 27 May, to concur in the petition and tell the Archbishop that he had put a stop to the distribution of the Declarations.[24]

Elsewhere, the reading of the Declaration did not imply support for the King or his policies. In the diocese of Hereford, Bishop Herbert Croft supported the reading of the Declaration and tried to encourage its reading, even though he was in favour of comprehension of Dissenters into the Church as a defence against popery. He was later angry that this action was interpreted as sympathy for the King's Catholicism.[25] Bishop Peter Mews of Winchester, who had been a natural supporter of James until the Magdalen College affair, was appalled by the Declaration. When Sunderland bypassed him and sent orders directly to the clergy of Winchester to read the Declaration, Mews countermanded the orders, telling Sunderland the he had done so 'conceiving that no one had anything to say in his diocese'.[26]

In the huge diocese of Chester, only three clergy read it and in the largest diocese, Norwich, only four did so.[27] None of the clergy in the city of Newcastle read the Declaration, and when he enacted a visitation there Bishop Crewe was received with little respect.[28] Isaac Archer, a low churchman in Suffolk, felt some trepidation at not reading the Declaration, in his prayers he 'begged respite of those dangers which hung over our heads for not reading the King's declaration of liberty'.[29] White Kennett wrote to Samuel Blackwell, on 3 June 1688, 'I am told the Declaration was read this day at Twyford, but omitted I believe in your Church and mine ... be pleased to remember me in your prayers'.[30] In contrast, Nathaniel Vincent, rector of Blo Norton in Norfolk, read the Declaration and also preached in Norwich Cathedral, on 29 May, that 'Rebellion is as the sin of Witchcraft'. It is certainly clear that some clergy felt they could not obey James because of the pressure from magistrates and landowners who were hostile to the King, and many of whom had been threatened with the loss of their offices on the grounds of their defence of religion; now they expected the clergy to do the same.[31] One estimate is that, nationally, only 400 out of 13,000 parish clergy read the Declaration.[32]

James's supporters on the bench of bishops tried to insist on the reading of the Declaration. Bishop Cartwright of Chester was zealous in insisting that his clergy should read the Declaration.[33] Lord Clarendon, was contemptuous of Cartwright calling him and Bishop Thomas Watson of St David's 'two scabby sheep' and 'very bad men [who] as they have no reputation or interest, so they are despised by those they court'.[34] Watson shared Cartwright's attitude to James II's Declarations of Indulgence.[35] As bishop of St David's, Watson made no secret of the warmth of his feelings for the King. He instructed the clergy of his diocese to read the second Declaration and attempted to force them to sign an address to the Crown; to those who refused he was said to have responded with 'menacing expressions' and accusations of disloyalty.[36] Watson also presented a congratulatory address to the Crown that had been wrung with considerable difficulty from the corporation of his home town of Hull. In a pamphlet calling on the clergy of St David's diocese to resist reading the Declaration, Watson was described as one of 'our temporising bishops, who under hand endeavour the ruin and destruction of the best of churches, like so many weathercocks turning every way with the fickle and inconstant wind'. Suspicions of Watson's papist sympathies were redoubled by his habit of ordaining and confirming with his hands crossed and, according to Morrice, he once drank to the pope's health. There were also rumours that Bishops White of Peterborough and Cartwright of Chester had privately been at fisticuffs over the Declaration.[37]

Some clergy were keen to explain publicly why they had read the Declaration. Edmund Ellis, rector of East Alling, Devon, published his reasons in *A Clergyman of the Church of England: His Vindication of Himself for Reading His Majesties Late Declaration*. Ellis explained that he did not like disobeying both his metropolitan and bishop; he had received both the Totnes apparitor with the King's Declaration and the Bishop of Exeter's letter asking him not to read the Declaration. In deciding whether he should obey the King or his bishop, Ellis had been influenced by St Gregory who had promulgated a decree of the Emperor though he had disagreed with it. Moreover, Ellis was convinced that, if the Declaration was not contradicted by Scripture or ecclesiastical canons, he would read it, and so he did.[38] Such public explanations of motives for reading the Declaration were relatively rare.

Lord Lonsdale believed that some clergy felt that if they complied with reading the Declaration, later something worse would be imposed on them but they would have lost their reputations with the gentry and landowners and therefore would be 'unpitied'. They 'could never take an opportunity of refusing upon a point more popular or justifiable; that their consenting to this made their conditions as precarious as that of anie other Dissenters, who having no legal establishment, were forced to flie to the Declaration for Protection'.[39]

On 6 June, a curious piece of propaganda against the bishops was published, 'with allowance' of the King, entitled *A Dialogue between the ArchB of C. and the Bishop of Heref. Containing the True reasons why the Bishops could not read the Declaration*. Bishop Croft of Hereford was a former Catholic, who had become fiercely anti-Catholic, but he was also a friend of James and a staunch supporter of the Church of England. It was only after nights of anxiety and indecision that he reluctantly permitted the Declaration to be read in his diocese. In the Dialogue, Sancroft was depicted rebuking Croft for abandoning his brother bishops, despite Croft's claim that the Church taught him to obey the King. But Sancroft had no answer to Croft's claim that the bishops were ignoring the weight of Church teaching; his answer was a feeble 'we are for the Dispensing Power...always provided it be on our side'. Sancroft was portrayed as an enemy to Dissenters, irritated by James's Catholic policies, and a demagogue.[40]

The second confrontation

Almost three weeks elapsed after his interview with them before James took any step against the bishops. During this period, James called sympathetic bishops to Whitehall and it was known he was consulting widely about the petition.[41] Later James was to complain that he had been forced into prosecuting the bishops because they had waited until 18 May before they presented their petition, but the time he took to decide on the next course of action gives away the lie to this claim.[42] It is clear that James wanted to see whether the Declaration was or was not widely read before he acted against the bishops. Jeffreys and Sunderland eventually advised James against further action, but by then James had gone too far. Moreover, the people of London were inflamed by the publication of the petition, and the King, regarding it as a breach of confidence, was persuaded to summon the Archbishop and the other bishops to appear before the Privy Council on 8 June, to answer the charge of misdemeanour. In the interim, the petition had been signed by five other bishops, who had not arrived in time to subscribe with the other prelates, but now added their signatures. Sir John Reresby saw the bishops on the day they met the King and recalled 'they all looked very cheerful and the Bishop of Chichester called to me asking how I did'.[43]

The role of Lord Chancellor Jeffreys in the matter is worth considering. Jeffreys had become lord chancellor as a reward for his brutal suppression of the West Country after the Monmouth Rebellion. But Jeffreys was a staunch Anglican and a close friend of Sancroft, Lloyd and White. He had tipped off Sancroft that if he did not allow Presbyterians to use the old charterhouse at Canterbury 'others' might get it, by which he meant the King's Catholic friends. Sancroft had advanced Jeffreys's brother and proposed him for a bishopric, but this was at a time when he had lost James's confidence and the proposal was rejected. Jeffreys was keen to avoid a confrontation with the bishops. But his personal relationship with Sancroft acted against the Archbishop's interest. To avoid sitting in judgement on his friend in the Ecclesiastical Commission, to which the petition could have been referred, Jeffreys advised the King to prosecute the bishops in the King's Bench.[44]

On Friday 8 June, at five o'clock in the afternoon James held a meeting of the Privy Council, and Sancroft and the six bishops were summoned to it.[45] The tone of the meeting showed the degree to which all trust between them had collapsed. So did the fact that Sancroft and the bishops had received detailed legal advice from Sir Robert Sawyer on the procedure of the Privy Council, and on whether they should incriminate themselves and offer recognisances for bail.[46]

'The King received them graciously', wrote Sancroft in his manuscript narrative of the scene, and the Lord Chancellor took a paper then lying on the table and, showing it to the Archbishop, asked,

'Is this the petition that was written and signed by your grace, and which these bishops presented to his Majesty?' Sancroft took the paper and, ignoring Jeffreys's query, said to the King, 'Sir, I am called hither as a criminal, which I never was before in my life, and little thought I ever should be, especially before Your Majesty; but since it is my unhappiness to be so at this time, I hope Your Majesty will not be offended that I am cautious of answering questions. No man is obliged to answer questions that may tend to the accusing of himself.' Provoked by this implied distrust, James exploded: 'Why this is downright chicanery! I hope you do not deny your own hand'.

'Sir', said the Archbishop, 'though we are not obliged to give any answer to this question, yet, if Your Majesty lays your command upon us, we shall answer it in trust upon Your Majesty's justice and generosity that we shall not suffer for our obedience, as we must if our answer should be brought in evidence against us.'

Lloyd agreed: 'All divined are agreed in this, that no man in our circumstances is obliged to answer any such question.'

'No', said James, 'I will not command you. If you will deny your own hands, I know not what to say to you.'

Lord Chancellor Jeffreys then asked the bishops to withdraw. In a few minutes they were called back again. The King then resumed the questioning of the bishops, saying: 'Is this your petition?'

The bishops responded: 'Pray, sir, give us leave to see it; and if, upon perusal, it appears to be the same...' and, after looking at it, they said 'Yes, sir, this is our petition and these are our subscriptions.'

'Who were present at the forming of it?' asked James.

'All we who have subscribed it' they answered.

'Were other persons present?' asked James.

'It is our great infelicity, that we are here as criminals; and Your Majesty is so just and generous, that you will not require us to accuse either ourselves or others.'

James gave up this line of questioning and asked one of the bishops: 'Upon what occasion came you to London?'

'I received an intimation from the Archbishop that my advice and assistance was required in the affairs of the Church.'

'What were the affairs which you consulted of?' asked James.

'The matter of the Petition.'

'What is the temper you are ready to come to with the Dissenters?' asked James, presumably intending to let the bishops know that he was well aware of their talks with the Dissenters.

'We refer ourselves to the Petition' was the reply.
'What mean you by the dispensing power being declared illegal in Parliament?' asked James, changing tack.
'The words are so plain that we cannot use plainer' replied one of the bishops.
'What want of prudence or honour is there in obeying the King?' asked James.
'What is against our conscience is against prudence, and honour too, especially in persons of character.'
'Why is it against your conscience?' asked the King.
'Because our consciences oblige us (as far as we are able) to preserve our laws and religion according to the Reformation' replied the bishops.
'Is the dispensing power then against the law?' asked James.
'We refer ourselves to the Petition' said the bishops.
'How could the distributing and reading the Declaration make you parties to it?' – meaning make the bishops party to the dispensing power.
'We refer ourselves to our petition, whether the common and reasonable construction of mankind would not make it so.'
'Did you disperse a printed letter in the country, or otherwise dissuade any of the clergy from reading it?' asked James.
'If this be one of the articles of misdemeanour against us, we desire to answer it with the rest,' replied the bishops.
In conclusion, the bishops acknowledged the petition and asked to receive a copy of any charges against them but they said that they asked permission not to answer further questions until they had been charged.
The bishops withdrew a second time, and then Jeffreys informed them that the King intended to proceed against them for publishing a seditious libel, and it would be heard before the King's Bench in Westminster Hall. James then made his worst error. As a calculated insult, he required them to enter into recognisances for their release prior to their appearance at the trial. Sancroft refused to do so, claiming his privilege as a spiritual peer made it unnecessary to do so. Before their meeting, Lord Clarendon had deliberately asked them what they intended to do if they were required to find bail. So the bishops had already discussed and planned for this eventuality. Sir Robert Sawyer had been asked to advise what they should do. Sawyer, who had been attorney-general for seven years until his

dismissal only six months previously because he would not support the dispensing power, advised them not to offer to find bail, and the bishops took his advice.

The King was clearly astonished and taken aback at the bishops' refusal to provide bail to secure their own liberty. He told them, 'I offer this as a favour, and I would not have you refuse it'.[47] Lloyd replied archly,

> whatsoever favour Your Majesty vouchsafes to offer to any person, you are pleased to leave it to him whether he will accept it or no; and you do not expect that he should accept it to his own prejudice. We conceive that this entering into recognizance may be prejudicial to us; and therefore we hope Your Majesty will not be offended at our declining it.

The King refused to consider allowing them to be released without recognisance. The bishops again withdrew from the council chamber.

The bishops were eventually joined by Lord Berkeley from the council chamber, who tried again to persuade Sancroft and the other bishops to enter into the recognisances; but they were immovable; so Berkeley returned to the council, and in about half an hour the sergeant-at-arms came out with a warrant to arrest them, and take them to the Tower.[48] He had another warrant addressed to the lieutenant of the Tower, commanding him to imprison them till their trial. Ironically, the lieutenant of the Tower was Sir Edward Hales, the Catholic whose case had founded the legality of the dispensing power.

To the tower

When the London citizens, who were in an excited state, thronging the areas around the palace at Whitehall to await outcome of the summons of the seven bishops to the Privy Council, saw them led out as prisoners under a guard of soldiers, and marched via Whitehall stairs to be taken by royal barge to the Tower, they were astonished and chased after the soldiers. Even the troops guarding the bishops knelt and asked for their blessings. The Archbishop, whose charities and hospitality during ten years at Lambeth – and before that as dean of St Paul's – had endeared him to the London poor and had won the affection and respect of the citizens of the city, tried to calm their indignation. 'Wonderful was the concern of all the people for them', wrote Evelyn, 'infinite crowds of people on their knees, begging their blessings and prayers for them as they passed out of the barge'.[49] Sancroft and the other bishops entreated them 'to preserve their loyalty to their sovereign, for they were bound not only to fear God, but to honour the King'. Both banks of the Thames were lined with the bishops' supporters, many on their knees. Some prayed for the deliverance of the bishops. William Bankes wrote to Roger Kenyon that as the bishops were being taken to the Tower, they heard of a pardon granted to a soldier who had deserted his colours and who, in the face of the hangman's rope, had refused the entreaties of a Catholic priest to abandon the Church of England 'this steadiness of the man give[s] the bishops great satisfaction' he wrote.[50]

When they got into the barge that was to carry them to the Tower, it was only with some difficulty that the people were restrained from rushing into the water after them. The bishops were cheered from both banks as they proceeded down the river, and when they reached the Tower, and landed at the Traitors' Gate, they were received with honour, the garrison, officers and private soldiers, knelt and begged their blessing. Affection for the Church of England was the prevailing sentiment, and the seven bishops were regarded as its champions. The bishops arrived just at the start of evening service, and were permitted by the lieutenant of the Tower to attend. By an extraordinary coincidence the appointed reading for the day was 'I have heard thee in a time accepted, and in the day of salvation have I succoured thee. Behold, now is the accepted time; behold, now is the day of salvation.' The bishops were treated with the utmost respect by the lieutenant, and allowed to see anyone they pleased. It was not only the people of London who sought the blessing of the bishops; the Revd Samuel Hill of Somerset wrote to Bishop Turner that 'the prayers of those that are bearing the Cross have more than an ordinary virtue and reception with our Blessed Saviour'.[51]

One of the features of the case of the seven bishops was the effectiveness with which news spread of the events. Within a few days the bishops' imprisonment was celebrated in verse, including *The Confinement of The Seven Bishops:*

> The Bishops prisoners are, we tamely see,
> The Reverend Prelates forced to bow the knee
> To Anti-Christ: No Mighty Monarch know

Tho' we must pay to Caesar what we owe;
There is a Power Supreme, by which you Live,
Whose arm is longer, and Prerogative
Larger by far, than Yours, whose very word
Can blast your Hopes and turn your two-edged swords;
Can make this Titular Vice-gerent know,
Vertue, like Palm's Deprest, do higher grow.
Tho' Roab'd in all the Grandure of the State,
Courtiers like Radient stars about you wait,
Midst of your Glorious joys, when you put on,
That awful presence, which becomes a throne:
Belshazzer like, Three words upon a wall,
'Twill dash Your joys, and make your glory fall
His Holyness, that Patron of Strife,
Tho' he can grant pardon, cannot Life.
Arise then, mighty Sir, in God-like mean,
As of thy valour, let thy truth be clear
By Action; Let Your Promises appear,
Protect the Church, which brought you to the Crown;
You know 'tis great, and Honourable to own,
A kindness done; But to reward with Death,
The Happy Instruments, That gave you Breath,
Is mean; and might a Catholick Conscience sting,
To cut the Hand of that, Anoints You King.[52]

The bishops were visited the next day by large numbers of nobility and distinguished people – including Bishop Henry Compton. At last the situation dawned on Jeffreys, who told Clarendon 'he was much troubled at their persecution'.[53] During their imprisonment, ten leading Dissenters visited the bishops, and assured them of their support. Robert Barclay visited them 'to justify a statement of which they had complained, that they had been the cause of the death of Quakers, but to assure them that the statement should not be used to raise prejudice against them'. To the nineteenth-century evangelical John Ryle, this was a debt of honour: 'at this critical juncture the Nonconformists, to their eternal honour, came forward and cut the knot... The shrewd sons of the good old Puritans saw clearly what James meant'.[54] Defoe said that the popular mood was that people would 'rather the Church of England should pull off our clothes by fines and forfeitures, than the Papists should fall both upon the Church and the Dissenters, and pull our skins off by fire and faggott'.[55] As Michael Barone has argued, 'the same Englishmen who were ready to believe Titus Oates's falsehoods about a Popish plot were ready, a decade later, to see a policy of alleged tolerance as an attempt to impose Popish tyranny'.[56] Oliver Heywood commented, 'though the Dissenters had liberty promised, we knew it was not out of love for us, but for another purpose. We heard the King had said he was forced to grant liberty at present to those whom his soul abhorred'.[57]

James, who felt his Declaration had been partly for the benefit of Protestant Dissenters, was disappointed and annoyed by their support for the bishops and summoned four of them. When he railed at their ingratitude, they replied 'that they could not but adhere to the prisoners, as men constant and firm to the Protestant faith'.[58] Worse still, Sir Edward Hales, lieutenant of the Tower, found he could not instil discipline among his troops to prevent them toasting the bishops' health. The impact of the bishops' imprisonment was in some case quite unexpected. Six years later, a Catholic priest conceded that 'when the seven bishops were sent to the Tower I changed my religion ... I recanted before Mr Tenison'.[59]

Bishop Compton, who visited the bishops regularly, received expressions of concern for the bishops from William, to which he replied that this concern 'had its just effect upon them: for they are highly sensible of the great advantage both they and the Church have by the firmness of so powerful a friend'.[60] The ambassador of the Holy Roman Emperor reported that 'affairs move every day closer to a catastrophe ... and although the King seems to know that he is in trouble, he does not in the least give the impression that he has any intention of giving up his project'. William clearly saw that the imprisonment of the bishops was an important opportunity. Twice at this time Princess Mary and her former chaplain, Dr Stanley, contacted Sancroft. The bishops' imprisonment was regarded by William as the best chance in his favour that James's rashness had presented. William wrote to Bentinck with some prescience 'this affair of the bishops could carry matters promptly to extremes'.[61] He naturally hoped that it would create a desire for revenge in the bishops' minds and those of their supporters. Stanley addressed a letter to Sancroft, by command of William and Mary, expressing their admiration of him and the other bishops:

> All men that love the Reformation do rejoice in it and thank God for it, as an act most resolute and every way becoming your places

(bishoprics we suppose he means). But especially our excellent prince and princess were well pleased with it (notwithstanding all the King's envoy here could say); they have both vindicated it before him, and given me command in their names to return your grace their hearty thanks for it, and at the same time to express their real concern for your grace and all your brethren, and for the good cause in which your grace is engaged. And your refusing to comply with King James II is by no means looked upon by them as tending to disparage the monarchy, for they reckon the monarchy to be undervalued by illegal actions. Indeed we have great reason to bless and thank God for their Highnesses steadiness in so good a cause.[62]

But Sancroft did not reply. There is no doubt that he was bowed down with sorrow, over the King's attack on the Church, but he did not anticipate help from a Calvinist foreigner.

5
The Trial

The bishops arraigned

Before the trial, the bishops had engaged the most able and astute lawyers of the day, some of whom had previously held office under James. These included Sir Robert Sawyer, the former attorney-general, Heneage Finch, the former solicitor-general and brother of Lord Nottingham, Sir Francis Pemberton (a former judge), Henry Pollexfen (who, as senior barrister on the Western Circuit had been obliged to prosecute the Monmouth rebels in the Bloody Assizes), Sir Cresswell Levinz, a former chief justice, Sir George Treby, Sir John Holt and John Somers. Pollexfen had insisted on the inclusion of Somers 'as the man who would take most pains and go deepest into all that depended on precedents and records'.[1] The bishops had at least three long meetings with their lawyers. The lawyers represented a wide range of political views including Pollexfen, Somers and Treby who were Tories. Cresswell Levinz was reputed to have been forced into the defence team with threats that his legal practice would dry up unless he accepted the brief to defend the bishops.[2] These were men who had figured prominently in the politically charged court cases of the past decade. In many cases they had expertise as judges and, in the case of Somers, of working closely with William Williams, the Solicitor-General. Lord Clarendon had advised the bishops on the best legal minds available, and they spent over 500 pounds on lawyers' fees. Numerous clerks were employed to copy documents; one spent all night copying documents for the trial.

In contrast, the King's lawyers, Thomas Powis, the Attorney-General, William Williams, the Solicitor-General and Serjeants Trinder and Baldock, were placemen of comparatively limited legal talents;

able lawyers such as Sir John Maynard having refused to act for the prosecution. Powis was said to want to do his best to be fair, though this was probably not what James was seeking. Williams, who had previously been a Whig but had become a Catholic, was described during the trial as 'violent and mighty zealous in the prosecution'.[3] Williams's careful preparation for the trial revealed the legal strategy. It is clear from Williams's brief that he had not anticipated that there would be any significant challenge to the dispensing power that could not be despatched by legal argument. The emphasis of his brief was to assemble evidence to 'prove' all of the contestable matters of fact. These were: the issuing of the King's first Declaration, the issuing of the second Declaration, the petition and its subscription by each of the bishops (here Williams had 20 witnesses who could attest that the signatures on the petition were indeed the signatures of the bishops themselves). Williams went on to marshal evidence that the bishops had presented the petition to the King. On the back of Williams's manuscript 'Information against the Archbishop of Canterbury and others...', he also scribbled issues that he might have to face in court. These included, for example, if the bishops claimed that a petition could not be a libel, that the dispensing power was illegal, and that the bishops were the King's spiritual judges and counsellors and therefore could advise the King without charge of libel. Williams's response to this was that if they were spiritual judges they should support the execution of the law.[4]

The imprisonment of the bishops lasted seven days; then they were released from the Tower by a writ of *habeas corpus*, from the King's Bench. Subsequently they could not walk in public without being mobbed by the people.[5] The release of the bishops reignited the public discussion of their treatment; the imperial ambassador wrote 'it is as yet impossible to tell what impression the unexpected imprisonment of these men will make. Time will show what will follow upon it, but this business must be considered as unavoidably leading to a great revolution'. Moreover from their imprisonment, they had been the subject of public prayers in Dissenting and Anglican churches.[6]

The hearing which released the bishops from the Tower took place on 15 June. Westminster and all its approaches were thronged with spectators; as one account had it 'an itch after things novel had occasioned a confluence of multitude of persons of all qualities, who were impatient to hear the success and events of the bishops'.[7] Lord Sunderland was hissed as he arrived, and the presence of Lord Danby, now openly regarded as the agent of William of Orange, 'struck terror into the hearts of the judges'.[8] When they arrived at Westminster, 'there was a lane of people from ... the waterside upon their knees as the bishops passed'.[9] The bishops, as they entered Westminster Hall, were cheered and accompanied by upwards of 60 supporters most of whom were peers. They were received with great respect by the court and permitted to sit, a grant without precedent in cases where the crown prosecuted. Twenty-one nobles were accommodated in the hall and this undoubtedly subtly influenced the proceedings.

The arraignment was presided over by Sir Robert Wright, the Lord Chief Justice, sitting with Mr Justice Holloway, Mr Justice Powell and Mr Justice Allybone.[10] The judges were safe pairs of hands in James's view. Wright had led the visitation of Magdalen College, Oxford; Allybone was a Catholic who sat as a judge under the dispensing power; Holloway had sat with Jeffreys in judgement on Algernon Sidney, only Powell was an unknown quantity having been silent during the *Godden v Hales* trial – which was assumed to be an assent to the judgement of the majority of judges.[11] The judges anticipated that all sorts of legal technicalities would be raised; indeed some felt that Wright hoped the case would be mired in legal technicalities so that it would avoid all mention of the dispensing power. At one point, as legal documents were read in Latin, the Bishop of Peterborough said, 'My Lord, We desire it may be read in English for we don't understand Law-Latin'. To which the Solicitor-General responded acidly 'No, my Lords the Bishops are very learned Men (we all know) pray read it in Latin'.

Central to the case, presented for the King by the Attorney-General and Solicitor-General, was James's Declaration and his right to assert it. So the Declaration was read to the court. Since the King's Declaration had been read into the record, the bishops' counsel insisted that their case should be also, and their petition was tabled. Despite the tedium of these proceedings, there were moments in which the bad feeling between the two sides was evident. At one point, when the Solicitor-General asked the court whether it was ready to hear the King's case, the Bishop of Peterborough whispered with his counsel Sir Robert Sawyer. The Solicitor-General spat out, 'My Lord, you had better look another way, and look towards the court, for there your business lies.'[12]

After the presentation of the King's case, the bishops were permitted to enter a plea as to why they should not be tried this was

> The said Archbishop and Bishops say, that they are Peers of this Kingdom of England and Lords of Parliament, and each of them is one of the Peers of this Kingdom of England, and a Lord of the Parliament, and that they being (as before is manifest) Peers of this

Kingdom of England, and Lords of Parliament, ought not to be compelled to answer instantly for the Misdemeanour.

The debate on the legal issue of whether they should plead took three hours.[13] Nevertheless, the Lord Chief Justice ruled that the trial should proceed and that the bishops should plead guilty or not guilty. Each bishop was asked in turn and pleaded not guilty. The Archbishop offered a paper to the court, containing a plea that they should be allowed sufficient time to prepare their defence. This request, though contrary to the practice of the court, was granted, and the Attorney-General indicated that their trial was scheduled for a fortnight's time. There only remained the issue of whether the bishops would be bailed or whether they would be returned to the Tower of London pending the trial. The Attorney-General had no intention of repeating the King's mistake of demanding that the bishops pay a recognisance. When asked if they were entitled to bail, he replied, 'they are baylable, no question of it, my Lord, if they please'. When, however, the Attorney sought bail of £1,000 for the Archbishop and £500 for each bishop, the Lord Chief Justice also realised that they were in danger of repeating the problem of the bishops refusing to pay the sums. Additionally, the Archbishop raised the question of whether a peer of parliament was able to be committed to prison for a misdemeanour before judgement.[14]

Eventually, the Attorney-General agreed to 'what sums the court pleases' and the Lord Chief Justice pondered the appropriate level of bail for the bishops. The court, keen to avoid James's error, initially granted them bail on their own recognisances, which they did not refuse but in the end they agreed to be bound to appear under a penalty of £200 for Sancroft, and £100 each for the bishops, and were released. Twenty-one peers came forward to pay their bail, and a rich London Dissenter begged the privilege of standing bail for Bishop Ken.[15] Then the bishops were permitted to return to their own homes. They were received by the crowd outside with rapturous acclamations, bonfires were made in the streets at night, and enthusiastic demonstrations of popular rejoicing continued till morning. To the chagrin of the Catholic staff at the Tower, their prisoners did not return to the Tower that night.[16]

The bishops released

On his release, Bishop Lloyd was taken to the home of Lord Clarendon, who marvelled at how the crowds sought to touch the Bishop's robe and kiss his hands. So large were the crowds that their carriage had to find a circuitous route to Clarendon's house.[17] Sunderland, in despondency, saw that the citizenry of London mistook the bail for an acquittal and built bonfires, only to find their hopes disappointed. Feelings ran strongly for the bishops. The rector of Stoke Newington, Sidrack Simpson, who had read the King's Declaration, wrote to apologise to Bishop Compton and said he had only read it because he had felt intimidated. More importantly, Dissenters promised Sancroft that they would not profit from the crisis by accepting any livings or offices if James offered them.[18] The bishops standing as heroes of Protestantism brought the majority of Dissenters onto the 'Anglican side'.[19]

As the hearing on the charges of seditious libel approached, the authorities became anxious. Jeffreys bitterly regretted the decision to prosecute the bishops, not least because he regarded the judges as 'rogues', by which he meant that they were prone to react to political and public pressure.[20] On the night before the trial, James, exploiting his former friendship with Turner, made some attempts to persuade him to concede. But Turner was uncompromising, and stood firm in support of Sancroft and the others. James also had a private conversation with Sir Samuel Astry, who had the job of forming the jury, but to no effect.[21] The panel which comprised the jury was largely composed of Anglicans and leading businessmen and landowners.[22] There were a few Dissenters and a number of the jurors were employed by the King either in the revenue or in the navy, or were suppliers to the royal household. Most of them were magistrates, and several had court connections.[23] James was confident that a jury of this composition would find for him. Rumours were legion, some said that the King intended, if they were found guilty, to fine the bishops and devote the money to Roman Catholics. Others said that he would strip the Archbishop of his right to grant his *imprimatur* to printed books. It was also believed that the judges were handpicked for their servility to James.

During the summer of 1688, the work of the bishops' dioceses did not halt: Bishop Lloyd maintained a correspondence with Lord Herbert in June 1688 regarding the rectory of Newtown in his diocese. Despite Herbert's preference, Lloyd felt bold enough to appoint to the living the chaplain who had attended him in the Tower. He excused himself to Herbert, who had his own candidate for the rectory, by saying he did not know 'how His Majesty may deal with him' and therefore felt obliged to reward his chaplain at the first opportunity.[24]

Between the arraignment and the trial there was time for some others to enter into the debate on the issue. On 18 June, Lord Mountjoy wrote to Lord Dartmouth telling him that he was uneasy at the King's imprisonment of the bishops and offered his support for them.[25] On the same day, a *Letter from a Country Curate*... appeared. It was a crude piece of propaganda for James, masquerading as a defence of the bishops. The curate expressed support for the bishops in their petition but concluded that they had abandoned the doctrine of passive obedience and, by 'triumphing in the face of His Majesty', they endangered Protestantism. He concluded that the libel was the 'production of a traitorous head'.[26] A similar attack on the principles underpinning the bishops' position emerged in *Some Queries Humbly Offered to the Lord Archbishop of Canterbury and the Six Other Bishops Concerning the English Reformation and the Thirty Nine Articles*.

Nothing, least of all James's propaganda, seemed to diminish the support for the bishops however. William Wake wrote to friends in Oxford on 19 June:

> Perhaps nothing has given more strength and reputation to us with all sorts of men both at home and abroad than this brave stand, which they [the bishops] and the clergy have made. And I cannot but wonder if more do not argue, as an eminent Dissenter here lately did upon this occasion, that 'tis very odd that since His Majesty's principle is not to force conscience, our consciences should not be thought as dear to us and as worthy to be indulged as those who have not, I should think, such fair pretence to it.[27]

Brother bishops waited with baited breath, a few days later Bishop Levinz of Sodor and Man, brother of Cresswell Levinz, one of the bishops' counsel, wrote to Thomas Cholmondley:

> I have beene imployd in ye visiting my Lord Canterbury's diocese for him, else I had beene in your parts sooner, in order to my Voyage for my Island. I came back butt last night fro^m y^t imploy to Lambeth, & there I stay till tomorrow bee over to see w^t will become of ye Seven Champions of our Church, & then for Oxon.... You have always been a most generous patron to y^m, & I beseech y^u not now to forsake em, y^t so they may have all incouragemts to continue y^m firme to ye protestant church in this time of Tryall for I doubt n^t butt yu have heard wee have a Seminary of Romish Priests and Jesuits sending over thither which necessitates my going over wth all ye speed I can to withstand their pceedings to my uttermost & I hope God will give such a blessing to my weake

butt syncere indeavours, yt I shall preserve yt little nation from their corruptions, however I will doe my indeavours and leave ye event to God.[28]

The King's propagandists continued to be active as the trial approached. On 28 June, *Two Plain Words to the Clergy: Or an Admonition to Peace and Concord at this Juncture* was published by the King's authority under the text 'Who may say to a King, what dost thou?' In the heightened atmosphere of the prelude to the trial, and despite its objective, this tract was an extraordinary expression of provocative royalist hyperbole. Hindering the cause of repeal of the Test Act was claimed to be destroying 'the souls for whom Christ died and to make his precious death and blood of none effect'. The Test Act was 'the strong hold of Satan, from whence fiery darts are thrown at the blessed Jesus'. Much of the tract was an open attack on the Anglican clergy as 'lovers of your own selves, great boasters of your own Church-Learning... despising your Brethren'. They were accused of neglecting their flocks, and of them it was asked, 'have you not fleeced them?' It sought to persuade the clergy to 'account the King's favour and grace in granting the liberty... [and] Declare such things in your pulpits and prints'. James's motives were defended: 'we can have no reason to suspect that His gracious Promises are from ill or contrary ends'. But there were threats also: 'what do you think will be the consequence of not closing with the King in this great and glorious design?... Is the word of a King to be disbelieved? For the very shame! Was he not stiled James the Just?' The tract also targeted the Dissenters, seeking to shame them into action: 'You that are Dissenters, shew your selfes men: Men who have reason; you who have been traduced, reproached, rendered unworthy of the favour of your King'. In a vain attempt to call for quiet, it was argued 'we need not fear the Jesuiter and Covenanter, nor any of them, for we have a Prince [who] has promised to protect our Religion'. Above all, the clergy with 'persons also of all ranks and quality' should 'study to be quiet' and compliant to the King.[29]

Before the start of the trial of the seven bishops, a secret meeting was held at Lord Shrewsbury's house. The combination of the imminent birth of an heir and the trial of the bishops, had created a strong tide which had to be taken. In particular, the Queen's advanced pregnancy effectively snuffed out the strategy of sitting out James's reign until Mary could succeed and reverse everything. Five conspirators, Nottingham, Shrewsbury, Danby, Compton and Sydney, signed a letter with numerical ciphers which had previously been sent to William of Orange:

> This I suppose will be safely delivered, but yet I shall not say much: in a few days you will receive another, wherein you will know the minds of your friends. I believe you expected it before now but it could not be ready; this is only in the name of your principal friends which are 23, 25, 27, 31, 33 to desire you to defer making your complement till you have the letter I mention; what they are likely to advise in the next you might easily guess, and prepare yourself accordingly.[30]

The 'bed-pan baby'

Bishop William Lloyd was aware of much of the intelligence regarding the Dutch interest and the Queen's pregnancy. He wrote to Dr Dodwell, on 9 June, begging him for prayers for the Church but he also told Dodwell that regarding other matters 'you will be informed of by others'.[31] Lloyd was also actively employed in writing a pamphlet to discredit the birth of James's son, the 'bed-pan' baby, which happened on 10 June, two days after the committal of the seven bishops to the Tower. Lloyd was so appalled at the idea of the Princess of Orange being displaced as heiress presumptive by a Catholic heir apparent that he left no stone unturned to invalidate the claims of James's son. Lady Clarendon visited Lloyd during his imprisonment in the Tower, she was a gossip and the two discussed the scandals they had heard regarding the birth of the new Prince of Wales. Lady Clarendon collected a raft of whispers discrediting the parentage of the infant; Lloyd strung these together in a pamphlet, which was too strong even for Gilbert Burnet.

Jeffreys and Sunderland floated the idea that the birth of the heir was an opportune moment for the King to issue a general pardon which could include the bishops.[32] But James, in chagrin and triumph, refused to consider it. In fact, on the night of the birth of the Prince of Wales, James's position hardened. Whereas formerly the bishops had been charged with 'scandalous libel', now James reissued the charge to that of 'seditious libel', a far more serious charge. James was both confident and vindictive.[33]

After the trial, in October, James called together those who had been present at the birth of his son and demanded that they attest to his legitimacy. Bishop Lloyd was very uneasy when he learned of

James's intention of summoning all the witnesses to the birth to make them swear to the reality of it before the Privy Council. Lloyd tried to confound the King's plan by persuading Sancroft not to go to the court. Lloyd even grasped at the straw of telling Sancroft that, as a senior peer, he might be called on to judge the fact of the Prince's birth and he could not act impartially if he had attested to the legitimacy of the birth.[34] Lloyd wrote to him the following on 26 October:

> I was told the last night as a secret that his Majesty intends to send for all the lords that were present at the examination of witnesses concerning the prince's birth, and to require them to subsign the examinations. This is agreeable enough with that which is printed in the *Gazette*, viz., that a full and particular relation of this matter will be made public. For the hands of all that were present will add very much to the authority of the relation. I need not say what it will seem to import. Your grace has that to say for yourself which perhaps few others can say that were present. You did not hear a great part of what the witnesses said. If that will pass for a sufficient excuse, your grace has no cause to complain of the badness of your hearing. But surely it will be better for the public if such an excuse can be found as will suffice for all that were present, and if all could agree to give the same excuse. It should seem by the calling of you thither, that either there is, or there is like to be, a dispute concerning the birth of this child; and whensoever that matter comes to be tried you are like to be judges. But if the judges are called to set their hands to an examination of witnesses ex parte, before the cause comes to be heard, it is a strange kind of preoccupation that will make all the world of the plaintiff's side, and be rather a prejudice than an advantage to the cause. I hope his Majesty will be aware of this, and will therefore spare you this unnecessary trouble. Howsoever I thought it a part of my duty to let your grace know what I have heard.[35]

Lloyd's letter failed to prevent Sancroft from attending the council, but it did reveal Lloyd's determination to discredit the birth of James's son, and therefore perhaps shows how much Lloyd had committed his support to William of Orange. Sancroft knew it was his duty to attend the council and hear the evidence of the birth of the Prince of Wales. In fact, Sancroft was convinced by the testimony of the witnesses, both Roman Catholic and Protestant, that the claim of a false 'bed-pan' baby was the fiction of political agitators.

Nevertheless, once begun, popular belief in the illegitimacy of the Prince was widely commented on, including in verse:

> But whereas Bishops, for your loyalty,
> He makes you great, he did bestow on ye
> To keep you safe, his greatest strongest fort
> While you were there the Tower was the court
> All fled from James, to you for blessings came
> Imprisonment immortalised your name;
> Bishops of England's Church are men of fame
> And since his dire design in law has failed
> He seems to smile, you are to Council call'd
> To hear the worthy loyal servants swear
> That at the birth of Wales's prince they were.[36]

Another broadsheet ballad poem asked vulgarly how the Queen had become pregnant when 'not all the stallions in Europe could have got her with child'. It was also suggested that the birth could easily have been proven legitimate if the King had allowed a bishop to be present, but this was obnoxious to him.[37] In Wales, it was reported that a print had circulated showing the Queen with a cushion where her pregnancy should have been.[38]

The trial in the King's Bench

The trial of the bishops was held on 29 June 1688. There was again a large number of lords present. The Lord Chief Justice had excluded law students from Westminster Hall to make more room for the nobles. The sight of such a large group of nobles 'indeed frited the Judges and the Jury for they fancied that every one brot (sic) a halter in his pocket'.[39] It was perhaps foreseeable that bringing together such a large number of nobles, many of whom were disenchanted with James, or had been evicted from their offices as lords lieutenant, gave them an opportunity to consider their view of events and discuss the future. The most disaffected of the grandees used the opportunity to canvass opinion among their peers, Sidney was especially active in such consultations.[40] The official account listed the following peers at the trial: the Marquises of Halifax and Worcester; the Earls of Shrewsbury, Kent, Bedford, Pembroke, Dorset, Bullingbrook, Manchester, Rivers, Stamford, Carnarvon, Chesterfield, Scarsdale, Clarendon, Danby, Sussex,

Radnor, Nottingham and Abington; Viscount Fauconberge, and Lords Newport,[41] Grey of Ruthyn, Paget, Chandoys, Vaughan of Carbery, Lumley, Carteret and Ossulston. The record concluded 'Tis possible more of the peers might be present...whose names, by reason of the Croud, could not be taken.'

After the preliminaries, the defendants answered that they were present. The jurors, led by Sir Roger Langley, were sworn that they would 'well and truly try this Issue between our Sovereign Lord the King, and William Lord Archbishop of Canterbury, and others, according to your Evidence, so help you God'. The charge was read:

> That [the bishops]...did consult and conspire among themselves to diminish the Regal Authority & Royal Prerogative, Power and Government of our said Lord the King in the premises, and to infringe and clude the said Order; and in prosecution and execution of the Conspiracy aforesaid, They...with Force and Arms, &c. at Westminster aforesaid in the County of Middlesex aforesaid, falsly, unlawfully, maliciously, seditiously, and scandalously, did frame, compose, and write, and caused to be framed, composed, and written, a certain false, feigned, malicious, pernicious, and seditious Libel in writing, concerning our said Lord the King, and his Royal Declaration and Order aforesaid, (under pretence of a Petition) and the same false, feigned, malicious, pernicious, and seditious Libel, by them...with their own hands respectively being subscribed, on the day and year and in the place last mentioned, in the presence of our said Lord the King, with Force and Arms, &c. did publish, and cause to be published; in which said false, feigned, malicious, pernicious and seditious Libel is contained.

After the charge, the Attorney-General began with some opening remarks. He said that the prelates were not prosecuted as bishops, nor for any point of religion, but they were prosecuted as subjects of the King and for a temporal crime. Nor were they prosecuted for omitting to do anything, but for censuring the King and his government, and for giving their opinion relating to law and government. Powis claimed that no man was allowed to accuse even the most inferior magistrate of any misbehaviour in his office, unless it be in a legal course. To say that they would not obey the King was 'against the Rules and Law of the Kingdom'. He accepted that men would always claim that governments got things wrong, but there were proper means for remedying grievances which the bishops had not adopted. The Attorney-General also took the opportunity to remind the jury that

> His Majesty, of his great Clemency and Goodness to his People, and out of his desire that all his Subjects might live easily under him (of which I think never Prince gave greater or more plain Evidence of his Intentions that way) the fourth of April, 1687. He did issue forth his Royal Declaration for Liberty of Conscience; this matter, without all question, was welcome to all his People that stood in need of it, and those that did not, could not but say, the thing in the nature of it was very Just, and Gracious.

He also argued that it was reasonable for the King to ask the bishops to read the Declaration in the churches of their diocese and that the King was rightly aggrieved that they did not.

The court heard, at considerable length, the formalities of the issuing of the Declaration under seal, which was designed to emphasise the legality of the Declaration. Indeed a copy of the Declaration was presented, so that the Lord Chief Justice could see it had been issued under the Great Seal. It was handed to Sir Robert Sawyer so that he could see the Great Seal also. There was also great play of the fact that the Declaration had been printed by the King's order. The laborious proving of facts continued with the testimony that the signatures on the bishops' petition were in fact the signatures of the archbishop and bishops. At one point the Lord Chief Justice impatiently urged a witness to give proof that he knew of the bishops' signatures. The Bishop of Peterborough intervened and addressed the Lord Chief Justice saying

> My Lord, we are here as Criminals before your Lordship, and we are prosecuted with great Zeal: I beg your Lordship that you will not be of Councel against us, to direct 'em what Evidence they shall give.

The Lord Chief Justice replied

> My Lord of Peterburgh, I hope I have not behaved my self any otherwise hitherto than as becomes me: I was saying this (and I think I said it for your Lordships advantage) That this was not sufficient

Proof; and I think, if your Lordship observed what I said, it was for you, and not against you.

Serjeant Pemberton retorted 'Pray, my Lord of Peterburgh, sit down, you'll have no wrong done you'.

There were other examples of bad feeling during the trial. When a number of the lawyers spoke at once, the Lord Chief Justice said 'Let one of you ask a question at a time, and not chop in one upon another.' Later, the Solicitor-General complained 'In all the Tryalls that ever I have been in, in all the Cases of Criminals, the King's Witnesses used to be treated with respect, and not to be fallen upon in this manner.' On another occasion, when the Solicitor-General complained that a question was ill-mannered, he told the bishops' lawyers that he was noting who asked which question, to which Serjeant Pemberton replied 'record what you will, I am not afraid of you, Mr Solicitor!' Sir John Reresby commented 'the King's counsel [were] so much outdone that it was wished at court that the thing had never been begun'.[42]

There were further tedious technical issues, including whether the charge was correct in claiming that the bishops had published their petition in Middlesex. At one point, the defence lawyer Creswell Levinz nearly secured their early acquittal by taking the objection that there was no evidence of their petition being published in Middlesex. Sunderland was forced to scuttle round and find the evidence and get it to Westminster Hall. When he returned, there was a moment of comedy; Sunderland was hissed by the crowd as he stepped from his sedan chair, and a voice shouted 'Popish dog!' A man who bowed to Sunderland was kicked hard from behind.[43] The bishops conceded that they signed it at Lambeth – which was in Surrey – but denied that they had printed it and circulated in the streets of London. William Bridgman of the Privy Council claimed that

> within a few hours after I saw it, the King shewed it to several people, and he said, it was the Petition the Bishops had delivered, he took it into his own custody, and afterwards commanded me to write a Copy of it, and there was no Copy made of it but that one, but notwithstanding that, I do remember I did see a Copy of the Petition, within a day or two after it was presented about the Town.

Sunderland was mute on whether he had circulated the petition, simply confirming that the bishops had come to him, asked him to read the petition and that he arranged for them to see the King.

In his statement for the defence, Sir Robert Sawyer claimed that the petition

> only contains their Reasons, whereby they would satisfy his Majesty why they cannot comply in a Concurrence with his Majesty's Pleasure; and therefore they humbly beseech the King, and beg and request him (as the words of it are) that his Majesty would be pleased not to insist upon their distributing and reading of this Declaration.

He told the jury:

> there is nothing in this Petition that contains any thing of Sedition in it..., I appeal to your Lordship, the court, and the Jury, whether there be any one word in it, that any way touches the King's Prerogative, or any title of Evidence that has been given to make good the Charge; It is an Excuse barely for their non-Complyance with the King's Order, and a begging of the King with all Humility and Submission, that he would be pleased not to insist upon the reading of his Majesty's Declaration upon these grounds, because the Dispensing Power upon which it was founded, had been several times in Parliament declared to be against Law, and because it was a Case of that Consequence that they could not in Prudence, Honour or Conscience concur in it.

The issue then turned to whether the bishops had challenged the King's claim to have a dispensing power. Wright tried to prevent any discussion of this but was prevented from doing so by Justice Powell who asserted that this was the very crux of the case.[44] There had been divisions between the bishops and their lawyers as to whether to make such a frontal assault on the King's prerogative. Nottingham, who was gradually abandoning his caution, advised that the bishops' counsel should make a clear challenge to James's dispensing power.[45] How to make the challenge was the idea of John Somers and it took the prosecution by surprise. Sawyer addressed the court very simply:

> My Lords the Bishops were commanded to do an Act, which they conceived to be against Law, and they decline it, and tell the King the reason; and they have done it in the most humble manner that could be, by way of Petition. If they had done (as the Civil Law terms it) Rescribere generally, that had been lawful; but here they have

done it in a more respectful manner, by an humble Petition. If they had said the Law was otherwise, that sure had been no Fault; but they do not so much as that, but they only say, it was so declared in Parliament; and they declare it with all Humility and Dutifulness ... The Question is, Whether they are guilty of Contriving to diminish the King's Regal Authority, and Royal Prerogative, in his Power and Government, in setting forth this Declaration? Whether they are guilty of the making and presenting a malicious, seditious and scandalous Libel; and whether they have published it, as it is said in the Information, in the King's Presence?

Now, my Lord, Where is the Contrivance to diminish the King's Regal Authority, and Royal Prerogative? This is a Declaration founded upon a Power of Dispensing, which undertakes to suspend all Laws Ecclesiastical whatsoever; for not Coming to Church, or not Receiving the Sacrament, or any other Nonconformity to the Religion established, or for or by reason of the Exercise of Religion in any manner whatsoever; Ordering that the Execution of all those Laws be immediately suspended, and they are thereby declared to be suspended; as if the King had a Power to suspend at once all the Laws relating to the establish'd Religion, and all the Laws that were made for the Security of our Reformation. These are all suspended by His Majesty's Declaration (as it is said) in the Information, by virtue of his Royal Prerogative, and Power so to do. Now, my Lord, I have always taken it, with Submission, that a Power to abrogate Laws, is as much a part of the Legislature, as a Power to make Laws: A Power to lay Laws asleep, and to suspend Laws, is equal to a Power of Abrogating them; for they are no longer in Being, as Laws, while they are so laid asleep, or suspended: And to abrogate all at once, or to do it time after time, is the same thing; and both are equally parts of the Legislature. My Lord, In all the Education that I have had, in all the small Knowledge of the Laws that I could attain to, I could never yet hear of, or learn, that the Constitution of this Government in England was otherwise than thus, That the whole Legislative Power is in the King, Lords and Commons; the King, and his two Houses of Parliament. But then, If this Declaration be founded upon a part of the Legislature, which must be by all Men acknowledged, not to reside in the King alone, but in the King, Lords and Commons, it cannot be a legal and true Power, or Prerogative.

This, my Lord, has been attempted, but in the last King's time; it never was pretended till then; and in that first Attempt, it was so far from being acknowledged, that it was taken notice of in Parliament, and declared against: So it was in the Years 1662 and 1672. In the Year '62 where there was but the least Umbrage given of such a Dispensing Power; although the King had declared, in his Speech to the Parliament, that he wished he had such a Power, which his Declaration before seemed to assume; the Parliament was so jealous of this, that they immediately made their Application to His Majesty, by an Address against the Declaration; and they give Reasons against it, in their Address: One, in particular, was, That the King could not dispense with those Laws, without an Act of Parliament.

Somers's brief for Wright went on to cite the 1674 case of *Thomas v Sorrel*, in which the judges had rejected the dispensing power unless sanctioned by Parliament.[46] Therefore, if the dispensing power was illegal, the charge of seditious libel against the bishops could not be true as there was no sedition and no libel.

Mr Pemberton defended his clients with the additional claim that they had done their duty to the King and to the Church. Mr Levinz went a little further by saying, 'If the King could dispense without an Act of Parliament, what need was there for the making of it?' Which strayed somewhat into the quagmire of constitutional theory. In fact, Levinz's point led to a legal wrangle over precedents for the dispensing power going back to the reign of Richard II, and the Solicitor-General even threatened to send for the rolls of Richard's reign. Levinz countered that he need not go back to Richard II when as recently as 1662 and 1672 the power had been denied to the King by Parliament. Indeed in 1662, the King had been recorded as saying that he 'could heartily wish I had the power of indulgence', and in 1672 the Commons had been as blunt as to tell the King 'that penal statutes in matters ecclesiastical cannot be suspended but by act of Parliament'.[47]

There was also some debate on the right of every subject to petition the King, with the implication that the prosecution implicitly impugned that right. Despite the fact that the seven bishops were not tried as peers in the House of Lords, it was unclear whether the bishops' status as spiritual peers affected their right to petition the King. The bishops argued that they had the right to petition the sovereign at any time, while the prosecution claimed that this right was only permissible when Parliament was in session (which, at the time of the delivery of the petition, it was not). If the bishops were only lords of Parliament, and not peers, their right to petition would be terminated while Parliament was dissolved. The bishops' lawyers argued that peers were counsellors of the sovereign, whether Parliament was in session or not; therefore, if the bishops were peers, they would be free to present petitions.

The King's lawyers argued that nothing better confirmed that the bishops had challenged the King's power than the crowd which had turned up to the court that day, since the willingness to dispute the King's power and authority had a destabilising effect. The bishops had an opportunity to raise their concerns in Parliament, and the King's intention to call a Parliament in the autumn was known to the bishops, so that they had deliberately ignored the opportunity to legally raise their grievance.

After these legal arguments, the Lord Chief Justice asked the Solicitor-General to 'come close to the business, for it is very late'. But the Solicitor-General had one further point, which was to respond to Mr Justice Holloway's question of what would he have had the bishops do to solve their consciences, especially given that the Parliament was only to be held in November. The response was that they should have acquiesced until the meeting of Parliament. The record noted 'At which some People in the court hissed'. The Attorney-General jumped up and said 'This is very fine, indeed; I hope the court, and the Jury, will take notice of this Carriage'.

The Lord Chief Justice said that he thought that the bishops were entitled to petition the King, and were not required only to do so through Parliament. But if they had the right to petition the King 'they ought to have done it after another manner'. If they petitioned in the way they had, it made the government 'very precarious'. As the trial came towards its close, Serjeant Baldock insisted on the right to address the court. He didn't deny that subjects had the right to petition the King, but they also had the duty to obey him. He refused to enter the discussion of whether the King possessed the dispensing power, or whether Parliament had denied the King that right. He simply advanced the view that the only thing the bishops were commanded to do was to order the Declaration to be read in their dioceses. He accused the bishops of asserting that the King had done an illegal thing, and this alone was a seditious libel. He concluded, 'I shall leave it here to your Lordship, and the Jury, whether they ought not to Answer for it.'

The Lord Chief Justice hoped the case would close with Baldock's comments, but Serjeant Trinder also demanded the right to speak. The Lord Chief Justice weakly protested 'How unreasonable is this now, that we must have so many Speeches at this time of Day? But we must hear it; go on Brother.' Trinder raised the red herring of whether, if the King's claim to dispensing power were illegal, the case would fall? The Lord Chief Justice brusquely retorted: 'That is quite out of the Case.'

In summation, the Lord Chief Justice briefly went back through the evidence. Some felt that the show of public support for the bishops, particularly among the peerage, had unsettled Wright. However, he concluded that the dispensing power was irrelevant, and that the only issue for the jury was whether the bishops' petition was in effect a publication of a seditious libel. This was where the Lord Chief Justice made a grave error. In the seventeenth and eighteenth centuries, as today, juries were only permitted to decide matters of fact, in this case whether the libel, the petition, had been published. It was a legal judgement, a matter for the judges to rule, whether the libel was seditious. But the Lord Chief Justice permitted the jury to decide on both issues. Here, James was badly served by his own placemen. The Lord Chief Justice explained that anything that disturbed the government could be termed a libel, and briefly he gave them his own view: 'I do take it to be a Libel.' Having indicated his view, the Lord Chief Justice invited his fellow judges to give their opinions. Mr Justice Holloway, noting how remarkable it was for anyone to give their view after the Lord Chief Justice, said that this was a remarkable trial. He briefly reviewed the case and concluded that it was the right of all subjects to petition the King and that the bishops were only guilty if they had done so with ill intent. But Holloway observed, 'I cannot think it is a Libel: It is left to you, Gentlemen, but that is my Opinion.' The Lord Chief Justice irritably retorted that he had not asked his fellow judges to comment on the trial, but just to give their legal opinion. Mr Justice Allybone said that no man could write against the exercise of government without committing a libel and that the King of England was an 'absolute sovereign'. Consequently, 'this, say I, is a Libel ...' Allybone's reputation had suffered more than many involved in the trial. Macaulay claimed he showed 'gross ignorance of law and history', and 'brought on him the contempt of all who heard him'. More recently, William Speck has claimed his comments were 'a perfectly valid summary of the law as it stood' in 1688.[48] Mr Justice Powell, who had been a pupil of Jeremy Taylor in Wales, astonished the court by arguing that the dispensing power the King claimed 'amounts to an abrogation and utter repeal of all laws', and that there was no difference between the dispensing power and abolishing parliament. He summed up simply, 'Truly, I cannot see, for my part, any thing of Sedition, or any other Crime, fixed upon these Reverend Fathers, my Lords the Bishops.' In 1689, Sir John Hawles, MP for Old Sarum, was to claim that the other judges' apparent support for the dispensing power was not advanced out of conviction: 'some did it out of weakness, others out of fear, and not out of the dictates of their judgement'.[49]

The exhausted Lord Chief Justice then committed the case to the jury and, as was customary, asked 'Gentlemen of the Jury, have you a mind to drink before you go?' Wine was sent for and the jury was asked whether they wanted to have any of the papers to review as evidence. They were given the statute book, but the Attorney-General tried to prevent them having the bishops' petition also. However, the Lord Chief Justice allowed the jury to have copies of the petition and the King's Declaration. The official record noted: 'Then the court arose, and the Jury went together to Consider of their Verdict, and stayed together all Night, without Fire or Candle.' So a major constitutional point was debated overnight in complete darkness, the jury were also denied bread, tobacco and drink. About midnight they were reported 'very loud out among one another'. At four in the morning, they called for basins of water and towels which they were allowed.[50] By some accounts, when the jury began deliberation it was split 7–5 for acquittal and gradually eroded the views of the five in favour of conviction.[51] Some rumours spread that the last juror to concede was a brewer dependent upon royal patronage, who held up the verdict until 6 a.m.[52] It is clear that the bishops, or at least some of them, expected to be convicted. Sancroft and Turner both prepared speeches for such an eventuality and Sancroft pointedly told Bishop Sprat, well-known for his love of luxury, that he knew how to survive on £60 a year.[53] Sancroft was also well-informed: his agent, Mr Ince, reported all the news of the jury to Sancroft, including their demand for water and towels at 4 o'clock, and the noise from the jury room, in a note written from the Bell Tavern in King Street at 6 o'clock that morning.[54]

The acquittal

The court reassembled the next day at nine o'clock. The jury responded to the question whether they had reached a verdict. They indicated they had and the foreman, Sir Robert Langley, delivered the verdict of not guilty. 'At which there were several great Shouts in the court, and throughout the Hall.' Clarendon recalled the cry was so loud 'one would have thought the Hall had cracked'. The Solicitor-General asked the Lord Chief Justice to take one of those rejoicing, a member of Grays Inn, into custody and he was reproved by the Lord Chief Justice. He then said 'Sir, I am as glad as you can be, that my Lords the Bishops are acquitted; but your Manner of rejoycing here in court is Indecent, you might rejoyce in your Chamber, or elsewhere, and not here.' Edward Clark wrote that the verdict was received with 'inexpressible joy of all Good Protestants, signified by their loud acclamations even in ye Hall sitting all the courts of Justice, which by the multitudes of the people attending was soe great that they were plainly heard through most parts of ye City'.[55] It was a verdict later claimed to the 'immortal honour' of the jury.[56]

The Marquis of Halifax, overcome with excitement, waved his hat over his head, cried 'Huzza!' He was later to claim that the bishops' trial had brought the people of England together 'and bound them with a knot that cannot easily be untied'.[57] The crowd took up the shout from him; it filled the hall, and was repeated by others waiting in Palace Yard and around Westminster Abbey. Sir John Reresby called it 'a little rebellion in noise though not in fact'.[58] Solicitor-General William Williams was hissed and Bishop Cartwright called a 'wolf in sheep's clothing'.[59] The news quickly spread through London. Surrounded by congratulating friends, the bishops left Westminster Hall. It happened to be St Peter's day, and the bells were chiming for morning prayers. Sancroft extricated himself from the crowd by entering the Chapel Royal at Whitehall with the six other bishops. Together, in the King's new Catholic chapel, they offered up their prayers and thanksgivings for their acquittal. By a happy coincidence, the New Testament reading for the day was from the twelfth chapter of the Book of Acts, recording St Peter's miraculous deliverance from prison. When Sancroft returned to Lambeth, the grenadiers from Lord Lichfield's regiment formed an escort, lining the route from the river to the entrance to the palace, each asking his blessing.[60] The news travelled so fast that a friend of Sir Charles Kemys, writing from London on the morning of the acquittal, did not include all the details as 'they could not fail to come to you', but he indicated that 'everybody here this morning is engaged in it [the acquittal] the one way or the other'.[61] As one anonymous author later wrote, 'the nation now began to be in high ferment'.[62] The rejoicing of the people continued all day, and into the night, during which Mr Coggs, beadle of the Duchy Liberty in London, was accidentally killed in the celebrations.[63]

That night, between Charing Cross and Somerset House alone, there were 56 bonfires, and they were often burnt in front of the homes of known Catholics. James, in exasperation, threatened to quarter troops in London.[64] The Lord Mayor did all he could to suppress bonfires, but in vain. He had had some warning since the London mob had lit bonfires when the bishops had been granted bail, assuming they had been acquitted.[65] Some illuminations were in the form of seven golden candlesticks, of which the tallest represented Sancroft, the other six represented the bishops. This was a clear allusion to the sacred candlesticks in the Book of Revelation.

The commercial classes took immediate advantage of the demand for episcopal memorabilia. The Huguenot engraver, Simon Gribelin, produced two engravings of the events. The first image represented the seven bishops as the seven churches of Asia Minor referred to in the Book of Revelation, and a quotation from Revelation was engraved on the image. A second engraving contained a portrait of each of the seven bishops on a candlestick. The allusion was clear: the bishops' persecution was part of the extraordinary events that would lead to the Apocalypse.[66] A silver medal was designed and struck, having a portrait of Archbishop Sancroft in the centre, and those of the six bishops associated with him in his imprisonment and trial grouped round him. Such was the demand for these medals that a cheaper, bronze, version was also struck. Before long, the broadsheet ballads that circulated in London were also celebrating the bishops. In *The Prince of Orange Welcome to London* the events leading up to the invasion were recounted including

> The Bishops away to the Tower was sent
> As stout and as cheerful as martyrs they went
> Not fearing what ever might fall to their doom
> They scorn to submit to the Clergy of Rome...
> Now while the true church thus did tottering stand
> It was a great grief to most men in this land
> But while we with sorrowful sighings did grieve
> Each fryar and Jesuits caught in their sleeve.[67]

About the same time appeared, *A Third Touch of the Times*, which recited:

> Nothing would serve you but Protestants ruin
> Rackings and Gallows and Knives and Fire
> You had provided, but 'twas your undoing
> Now you shall have them your selves for hire
> Yea so unbridled was your Power
> (tho' they no evil had deserv'd)
> You must go send the Bishops to Tower
> But now your rogueships shall worse be served.[68]

The same was the import of *A View of the Popish Plot*, which was decorated by a woodcut of a bishop burning at the stake, redolent of the reign of 'Bloody Mary'.[69]

It was not only in London that the people rejoiced, as news spread the rest of the country joined the celebrations. John Stevens wrote in his diary:

> The first hardened piece of insolence that I observed, was upon the news of the 7 Bishops being released, at which time in Welsh Poole in Montgomeryshire where I commonly resided, & many other places about, were made public bonfires in contempt of His Majesty's Proclamation forbidding the same, or rather in defiance of his authority.[70]

Bonfires were also lit at, among other places, Bedford, Bristol, Cambridge, Exeter, Gloucester, Lichfield, Norwich, Oxford, Salisbury and Tamworth.[71] Simon Patrick recorded

> I cannot but here remember with what joy the news of their being cleared was received at Peterborough. The bells rang out from three o'clock in the morning till night; when several bonfires were made...whereas the day before, which was a thanksgiving for the birth of the Prince of Wales, the bells did not stir till twelve o'clock.[72]

Similarly at St Giles in the Fields soon after the trial, when prayers were said for the royal family, the congregation fell silent and refused to read the prayers, instead they laughed and talked to each other.[73] In Scotland, where Anglicanism was not altogether well-regarded, Sir George Mackenzie, the Lord Advocate, wrote to Sancroft, 'it will doubtless be strange news to heare that the Bishops of England are in great veneration among the Presbyterians of Scotland, and I am glad that reason has retain'd so much of its old empire among men'.[74] Other leaders of the Scottish Presbyterian church – usually hostile to Anglicans – also wrote letters of support to the bishops.[75]

Across the country, grand juries refused to present or indict those who had celebrated the acquittal with bonfires, despite in Northamptonshire being subjected to a long diatribe to do so by the judge. At Winchester, when asked to present members of a mob for their part in the celebrations, the grand jury refused and instead presented the Catholic JP for holding office in violation of the Test Act.[76] Juries were emboldened on other religious matters; Middlesex magistrates found that the grand jury refused to return bills against Dissenters 'such kindness men generally had for the bishops and the Protestant religion'.[77]

At Oxford, only Magdalen College and Christ Church had bonfires to celebrate the birth of the heir, the other colleges remained silent.[78] But some people used the celebration of the birth as a covert way of toasting the bishops. On the news of the acquittal of the bishops, Arthur Charlett reported 'people in these parts will drink Bps Healths, the Bps Council and the Bps Jury'. In delight at the acquittal, a boy let a cat loose during the service at University College chapel, where the Catholic Obadiah Walker was Master.[79] At Peterborough, the bells rang all day and in Somerset effigies of the baby Prince of Wales were burnt.[80] Even High Church Tories, like Edmund Bohun, referred to the infant as 'the pretended Prince of Wales'.[81]

In Suffolk, Isaac Archer claimed that everyone rejoiced except the Catholics.[82] Certainly, the Dissenters seem to have been as relieved as the Anglicans.[83] Roger Morrice found that the acquittal of the bishops had changed his mind; it had shown the solidarity of Church and Dissent. As Mark Goldie claimed, 'Morrice displayed in microcosm the nation's painful realignments between Catholicism, Anglicanism and Dissent'.[84] In Warwickshire, there were celebrations in five towns and in Warwick the celebrations were led by the parish of St Mary's. Some clergy had felt the trial of the bishops very keenly. Lancelot Addison, Dean of Lichfield, felt himself also to have been on trial with the bishops because he felt so precisely their view of the Declaration.[85] In Lichfield, a 'very great ryot' greeted the news of the acquittal, the crowd celebrated with fireworks and the magistrates tried to disperse the crowd but were beaten with sticks.[86] In York, Sir John Reresby, the governor, reported that the streets were crammed with cheering people.

In Bristol, where Trelawny was bishop, 'all the bells rang, they gave very large expressions of joy and at night bonfires were made in many parts of this city', and the medals struck to celebrate the trial were distributed.[87] In Exeter, the news of the acquittal of the bishops, which only reached the city on 2 July, was met with bonfires, rejoicing and riots, and shouts of 'stand by the Church of England and Bishops!' The loyalist but equivocal bishop of Exeter, Thomas Lamplugh, had refused to sign the petition against the Declaration though he had not asked the clergy to read it, but Bishop Trelawny was widely admired in the West, consequently at the first sign of trouble Lamplugh fled the city.[88] From the evidence of the prosecution of rioters, it is clear that they included respectable citizens – goldsmiths, vinters and merchants – as well as the 'mob'. One of the centres of disturbance was the cathedral close, perhaps in recognition of the fact that the dean of Exeter had refused to read the King's Declaration in the cathedral. Ironically, the men responsible for suppressing the violence, and calling out the corporation fire engine, were the Dissenters who James had admitted to the corporation in November 1687 to replace the stubbornly non-compliant Tory corporation. In a strange reversal of fortunes, the city JP, Thomas Crispin, who had previously been prosecuted for holding Presbyterian conventicles now prosecuted Anglicans for their disturbance of the peace.[89] Further west in Cornwall, Trelawny's patrimony, the bells of Pelynt were rung and the mayor fired the town cannon, as did the mayor of Looe.[90]

On the evening of the acquittal, James, inspecting his army at Hounslow, heard the soldiers cheer. He asked what the cheer was and Lord Feversham replied 'nothing but the soldiers' joy at the bishops' acquittal'. James replied, 'you call that nothing?' But in public he was said to have commented 'Well! And what's that?'[91] He returned to London the next day, and outlawed public assemblies.[92] The only good news James got that month was Sunderland's stage-managed conversion to Catholicism, though he soon argued against the increasingly fraught attempt to get a Parliament to repeal the Test Act.

On 2 July, James authorised publication of *The Clergy's Late Carriage to the King Considered*. It railed against the Church for its disobedience but also expressed the King's bewilderment at the betrayal of the Church. 'The King was not only not obeyed by the clergy, where there was no sin in it, but where the obedience was purely ministerial. Had it been to renounce their own religion, or to receive His, it had been something...this is something surprising.' There was also confusion at the bishops' behaviour: 'this looks with an ill air, and carries too great a contradiction for men of their function and learning; and yet so it must be, or they are insincere in their petition.' The objection of the bishops and clergy was dismissed as 'a cavil and not a scruple'. The bishops had, it was concluded, created an artificial scare:

> they have acted, I mean their mock martyrdom, to force suffering and act it to a farce. What else can be their blessing people ten deep of a side with 'Have a ware of your religion', 'be faithful to your religion', 'the Lord strengthen you' etc and whilst not one tittle of their religion, but the liberty of other men's was the case: what shall an honest man think of this?[93]

The bishops were left with a significant legal bill, though the jury declined the traditional offer of 20 guineas for a meal, which would normally be charged to them. The fees of their legal team was nearly £250, but there was almost £150 in other expenses. The bishops taxed themselves at 6 per cent of their annual incomes to pay the bill, rather than force the poorest to pay as much as the wealthiest.[94] The total raised was over £600, of which £120 was spent on gratuities. Their solicitor received £48 and Somers, as junior counsel, received £32.[95] The bishops were also aware that they would be likely to be under threat from James. In some scribbled notes, which Archbishop Sancroft made for himself a few days after the acquittal, he suggested that they should 'establish a correspondence [with Dissenters]', he also speculated how the bishops should behave 'in case of a Popish visitation' and, most significantly, he laid down that 'the way of writing to the Archbishop is for every man to write to a private friend, and for him to deliver the letter to my Lord Archbishop'. A clandestine network of correspondents for the bishops was established. What is remarkable is the number of women who appeared in the list of secret correspondent addresses: Bishop Lloyd of St Asaph's contact was Lady Salisbury, Francis Turner of Ely's was Mrs Nalson or Madam Womock; Bishop Lake of Chichester's was Mrs Elizabeth Rowe; Bishop White of Peterborough's was a Mrs Clark. Possibly women's correspondence was less likely to be searched or read than men's.[96]

Short as the imprisonment and trial of Sancroft and the six bishops had been, it led to the most disastrous consequences for James, by producing an irreconcilable breach between him and the bishops. It was the more ill-judged on his part because it deprived the birth of his son of the most important and unimpeachable of witnesses, the Archbishop of Canterbury. If Sancroft had been present, and deposed that he was in the chamber when the prince was born, no one would have believed the later bed-pan stories. As it was, the Orange faction took every occasion to convert his enforced absence into circumstantial evidence that a spurious child had been imposed on the nation. But undoubtedly the most significant effect of the trial and acquittal was that it impeached James in the court of public opinion. There can have been few people who did not know of the trial, or find themselves taking sides – for King or Church. Whatever the euphoria of the moment of the acquittal, constitutional resistance to James had reached its limits, and executive authority remained in the hands of the King with his agenda largely unaffected.[97]

6
The Reaction

The stiffened resolve

Alfred Havighurst claimed that 'by implication they [the Jury] ruled against the legality of the suspending power as exercised by James in the Declaration of Indulgence'. He also argued that none of the judges in the trial of the seven bishops asserted the legality of the suspending power – the abrogation of a law generally – they asserted that the King might dispense with the law in certain specific cases. This was the distinction reflected in the Hales judgement. While the difference in legal terms is significant, in political terms it was a distinction lost on the jury and the public, and, during the trial, the terms were used interchangeably. Havighurst suggested that the pressure of public opinion swung the trial: 'in terms of popular pressure hardly any other trial in the seventeenth century is comparable'. During the trial, the public applauded arguments for the bishops and scorning those for the crown.[1] W. L. Sachse concluded that the acquittal was a verdict against James's whole system of government.[2] The lawyers at the trial had, in defending the bishops, developed general arguments against James which would be used later in the Revolution.[3]

Whatever the legal implications of the acquittal, Tim Harris has argued that it is wrong to see the expressions of support for the bishops as the same as endorsement of the Whig exclusionists' desire to be rid of James. Many of the supporters of the bishops, and a number of the bishops themselves, remained the traditional Tory Anglican adherents, who had supported James and the Stuart cause. Moreover, the leaders of the seven bishops had been well-known for their opposition to Dissent, and this may also have played a part in the expressions of support. Certainly the disturbances attendant on the trial were directed at the houses of London Dissenters who were collaborating with James.[4] It may be that, as time went on, the mob and the general momentum following the bishops' acquittal, became increasingly Whiggish. Certainly in Yorkshire, where Danby led opposition to James, the acquittal of the bishops was the signal for the start of tentative plans for open opposition to the King. A committee of Yorkshiremen was formed with district organisations. The local militia was brought up to full complement and men favourable to Danby's cause against James were made its officers. William's intentions seem to have been almost common knowledge in the county. One militia leader sent a report to Danby that the West Riding had three troops of horse and regiments of foot at readiness and a fourth in the city of York, and that they were exercising and training so as to be ready 'when summoned'.[5]

The claim of the bishops' imprisonment and trial to be a watershed in attitudes to James can be evidenced by the new-found willingness of others to resist him. During the trial, James received another rebuff. The chancellorship of Oxford had been vacant since the death of the Duke of Ormonde; in June 1688 James nominated Lord Chancellor Jeffreys to it. But the University chose Ormonde's grandson in public defiance of James.[6] The opposition of the bishops had clearly emboldened others. This was one of the principal achievements of the trial of the bishops. It was clear that James *could* be opposed and defied.

The most significant stiffening of resolve occasioned by the trial was the invitation to William, which was signed on the night on which the bishops were acquitted. This was no coincidence. The seven 'immortals' who signed the invitation were emboldened by the bishops' acquittal, and the widespread public displays of support it had elicited. Once it was signed, Admiral Herbert, disguised as a common sailor, left London for Holland with the invitation to William to come to England to save the Church and nation. The invitation had, in part, been drafted by John Somers, one of the bishops' defence lawyers, and hastily completed after the acquittal of the bishops.[7] The invitation read

> We have great satisfaction to find by 35,[8] and since by Mons. Zulestein that your Highness is ready and willing to give such assistance as they have related to us. We have great reason to believe we shall be every day in a worse condition than we are, and less able to defend ourselves, and therefore we do earnestly wish we might be so happy as to find a remedy before it is too late for us to contribute to our own deliverance ... We who subscribe this will not fail to attend your Highness upon your landing, and to do that lies in our power to

prepare others to be in as much danger in communicating an affair of this nature, till it be near the time of its being made public.[9]

It went on that 'nineteen parts of twenty of the people are desirous of a change'.[10] According to Carswell, 'such words could hardly have been written until the affair of the bishops had run its course'.[11] The invitation was partly motivated by the sense that the trial had created a momentum for change which needed to be exploited, and a fear that, if delayed, events could swing back to favour James. He might succeed in packing a Parliament, and placing the army entirely in the hands of Catholics. Among the seven signatories was Bishop Compton; the others were the Earl of Devonshire, Lord Danby, Henry Sidney, Lord Shrewsbury, Lord Lumley and Admiral Russell.

Two of the politicians who chose not to sign were Halifax and Nottingham. The former had promised to sign if the bishops were convicted and imprisoned. Nevertheless, after William's accession, Halifax wrote to him that the bishops' trial had united the Protestants against James.[12] Nottingham, though he admitted that Princess Mary had the right to protect her succession as the Protestant heir, felt that he ought not to sign, but agreed not to betray the plot.[13] Nottingham also wondered whether, in the light of the bishops' acquittal, it was necessary for William to come to invade England as he 'could not imagine that the Papists are able to make any further considerable progress'.[14] At the end of July, Compton wrote to William assuring him that the seven bishops would lay down their lives before they would depart from their commitment to defend the Church. Compton hinted, 'I should say something of myself, but I had so lately an opportunity of making my mind known to you that it can be no purpose to say more now to you'.[15] Thomas Tenison, an early confidante of the conspirators, dined with Simon Patrick on 8 August, told him of the letter, and advised Patrick to get his money out of London.[16]

When he received the letter, William was also told that 'one of the Bishops believed William had a just cause of making a war on the King'.[17] In fact, it seems likely that both Lloyd and Trelawny had reached this conclusion. But the motives of the conspirators were mixed. In retrospect, Charles Hornby wrote of the signatories of the invitation:

> The stomachs of these Babes were not able to digest the strong Doctrines of deposing Kings...The Prince was to come and put all things in order; and when the House was Swept clean, he it seems was to be set by, like a broom behind the door.[18]

Hornby was strictly accurate, what he said was also true of most of the seven bishops, particularly Sancroft, though perhaps not of Lloyd and Trelawny; but the signatories of the invitation were astute men who also knew that issuing the invitation had the *potential* to end in James's deposition.[19] Compton, Danby and Sidney probably realised that they could never trust James, others had not reached this point.

It seems clear that some of the bishops were quickly taken into the confidences of the 'immortal seven'. In August, Bishop White visited Mr Austin, one of the jurors who had acquitted the bishops. White found him in a melancholy mood and, when he asked why he was so depressed, Austin replied that things looked so dark and fearful. White replied 'what? Do you think the Prince of Orange will sit still and do nothing to help us?' Equally, when Nathan Wright, Lord Keeper of the Great Seal, visited Bishop Turner in Ely a short while later, he told Turner how much he rejoiced at his deliverance. Turner replied to Wright 'wait patiently awhile, and we shall all be delivered'.[20]

Some days after the acquittal of the seven bishops, Henry Sidney, William's leading English supporter, came to London on behalf of the Prince and Princess of Orange, ostensibly to congratulate the King on the birth of the Prince of Wales. In reality, Sidney came to find out the strength of support for the invitation. Among others, he found Lord and Lady Churchill, and their friend, Colonel Charles Trelawny, ready and willing to respond to William's call. Colonel Trelawny also answered for his brother, Jonathan. The trial had plunged Jonathan Trelawny deeper than ever into financial problems. He was still smarting from having to accept Bristol diocese, with an income of only £300 per annum, instead of Chichester, Peterborough, or even Exeter, for which he had hoped. It was joked that 'King James had sent seven of his bishops to the Tower to be tested, that five had been proved pure gold, but that Sir Jonathan of Bristol and Dr. Lloyd of St. Asaph had turned out only prince's metal'. Either way, Sidney found out that he could rely on the Bishop of Bristol.

William's agents in London also reported that, while James was making military preparations, some commanders in the army had formed an Association of Protestant Officers, among the footguards and cavalry regiments – the core of the army. The Association had been instigated by the colonels John Churchill, Charles Trelawny, Percy Kirke and Thomas Langston, commander of Princess Anne's regiment. These officers had previously attended meetings of the so-called Treason Club. All of them suspected that their places would soon go to Catholic officers and all had told Sidney they would respond to William's call. By 22 September, half a dozen captains in the Duke of Berwick's regiment laid down their commissions and were said to 'have spoken some dangerous words'.[21] James had created a similar disloyalty among naval officers when he appointed a Catholic, Roger Strickland, as admiral. Strickland permitted a Catholic mass to be said on deck, and sought to raise an 'abhorrence' – an address – from his captains against the acquittal of the seven bishops. But when 'one of them swore bloodily at him and threatened to beat him' there was a near-mutiny and James had to go to Portsmouth and quell it.[22] Nevertheless, James had ordered the army and navy to be brought to a state of readiness, and the seriousness of the situation still had not dawned on Sunderland, who told Barillon that Louis exaggerated the fears of an invasion. Moreover, Sunderland believed that the Dissenters were satisfied by James's concessions and 'the Church of England principles will keep them loyal'.[23]

After his acquittal, Bishop Lake returned to Chichester diocese where he was hailed as a hero. While touring his diocese, he was received by gentry and clergy with respect, and the Duke of Somerset gave a reception in his honour at Petworth. The same was true of Bishop Lloyd of St Asaph, who toured most of Wales promoting William's cause. Of his progress John Stevens wrote

> I perceiv'd that the Generality of the People began to be more open hearted than they had been when I left them, discovering their inclination to the Prince of Orange, & an aversion to the King. They had been work'd into this disposition by Dr Floyd [sic], then Bishop of St Asaph...famous for being one of the Seven Bishops who refus'd to read King James's Declaration of Liberty of Conscience, & no less for his enthusiastick pretensions to the spirit of Prophesy...This worth person as soon as releas'd from the Tower, to which he chose to be committed rather than give Bail for his Appearance, taking a progress through the country, preach'd at almost every Church & din'd or supp'd at the Houses of most gentlemen of any note where all the rest met, to incense the people against the King, and dispose them for what follow'd ... it was no small surprise to find so great an Alteration in so short a time.[24]

Bishop William Thomas of Worcester, who had refused to allow the Declaration to be distributed in his diocese, preached a sermon in his cathedral in August 1688, it was redolent of the issues raised by the trial of his brethren. He referred to the time as 'this critical age and climate' and it is difficult not to see some of his words as a comment on the times. In arguing that every human offence would bring divine judgement, Thomas claimed, 'the accusations are varied, yet Christ arraigns, before he dooms[25]: not like Augustus, precipitate sentence without the Traverse'.[26]

It was not only the great and good who experienced growing resolve in their dealing with James after the trial. In Poole, when James reissued the borough charter on 15 September 1688 the burgesses refused to accept the new charter because it contained 'many encroachments into their liberties'.[27] In Norwich, in August, the corporation refused the King's request that they admit a quaker to its membership.[28]

The pursuit of the clergy

Unwisely, James did not let the matter of the Declaration rest after the acquittal of the bishops. It was rumoured that he was proposing to prosecute the bishops' counsel and that he intended to charge the bishops themselves before the Ecclesiastical Commission.[29] These rumours came to nothing, but James was not repentant and pursued those clergy who had not read the Declaration. On 12 July, the Ecclesiastical Commission was ordered to find out, through diocesan chancellors and archdeacons, the names of those clergy who had refused to read the Declaration. Already James's episcopal supporters, Cartwright, Crewe and Watson, were punishing those clergy in their dioceses who they discovered had disobeyed the King.[30] By 24 July, the first archdeacons and chancellors returned their excuses for not supplying the names of the recalcitrant clergy; by mid-August almost all returns were made, most refusing to name the clergy. The only archdeacons who sent names were those of Durham, Lincoln, St David's, Buckingham and Chester. Most archdeacons claimed that they had no authority to find out the names of the clergy. Dr Thomas Wainwright, chancellor of Chester diocese, cleverly notified the Ecclesiastical Commission of the names of those clergy who *had* read the Declaration. A further demand for names was issued on 16 August.[31] Some archdeacons replied that there was no way of finding this out before they held their annual visitations, to which the Commission replied that names should be sent by 6 December.

The punishment of those who refused to read the Declaration was easier when the clergy were dependent on James for their offices. James ejected Dr Francis Hawkins from his post as chaplain in the Tower of London for failure to read the Declaration, though Hawkins refused to be turned out of his house in the Tower. On 18 August, James further showed his opinion by awarding the vacant diocese of Oxford to Timothy Hall, who had been bold enough to read the Declaration in London, but who it emerged had no doctoral degree and had once been a Dissenter. There followed a wrangle with Oxford University, which refused to award Hall a DD by royal diploma. In the end, Hall was not elected bishop by the canons of Christ Church, Oxford in an expression of defiance to James.

Sancroft's response to James's continuing campaign against those clergy who had refused to read the Declaration was to issue 'Some heads of things to be more fully insisted upon', which was distributed by bishops to all clergy. In it Sancroft reminded clergy of their vows and oaths, and urged them to strict observance of holy conversation, diligent catechising and preparation for confirmation. He asked clergy to maintain the decency of daily worship, to strictly observe the rubrics and urge people to frequent communion. This was a clear fight back against the activities of Catholic vicars general who were confirming Catholics and against the King and his judges.[32]

Sancroft's subsequent metropolitan articles to all bishops, issued on 16 July 1688, did not mince words. Indeed the tone of the preliminary letter was extraordinarily bold:

> By the contents of them [the articles] you will see that the Storm in which he [Sancroft] is does not frighten him from doing his duty; rather it awakens him to do it with so much more vigour: and indeed the Zeal that he expresses in these Articles both against the Corruptions of the Church of Rome on the one hand, and the unhappy differences that are among Protestants on the other, are such Apostolical things that all good men rejoice to see so great a prelate at the Head of our Church, who in this Critical Time has had the courage to do his duty in so signal a manner.[33]

The metropolitan articles themselves urged the clergy to catechise the children and youth – almost certainly with one eye on Leyburne and the Catholic initiative in mass confirmations.[34] They urged the clergy, especially 'in all market and other great towns, and even in villages and less populous places, [to] bring people to publick prayers as frequently as may be', the former perhaps a hint at the need for parliamentary boroughs to be buttressed for Anglicanism. But Sancroft did not only speak in generalities regarding the circumstances of the day. Article VII enjoined

> that in their sermons they teach and inform their people...that all Usup'd and Foreign Jurisdiction is for most just causes taken

away and abolish'd in this realm, and no manner of obedience or subjection [is] due to the same, or to any that pretend to act by virtue of it: but that the King's power being in his Dominions highest under God, they upon all occasions perswade the people to loyalty and obedience to his Majesty in all things Lawful, and to patient submission in the rest'

Thus Sancroft straddled passive obedience and resistance with the reference to 'in all things lawful'. But the next article encouraged clergy to 'maintain fair correspondence... with the gentry and persons of quality in their neighbourhood, as being deeply sensible what reasonable assistance and countenance this poor Church hath received from them in her necessities'. Sancroft was seeking to strengthen the alliance between the Church and gentry to defend the status quo. Article IX spoke of the need of the clergy to be vigilant to 'all seducers and especially of Popish emissaries, who are now in great numbers gone forth amongst them, and more busie and active than ever'. The articles urged that, since the Catholic clergy – 'evening wolves' – tended to 'unsettle' people at times of sickness, and at the end of their lives, the clergy should especially visit the old and infirm. The final article urged clergy to 'walk in wisdom' towards Protestant Dissenters and 'frequently to confer with them in the spirit of meekness' to win them to Anglicanism and to have 'a tender regard' for Dissenters. They were to be treated 'calmly and civilly' and promised that the Church of England was an 'irreconcilable enemy' to the Church of Rome and the divisions between Anglicans and Dissenters 'were altogether groundless'. This was a new Sancroft, not the primate of the Test Act and the Clarendon code.[35]

In the face of the polarised positions of the King and bishops, James's natural supporters were weakening in their support. Bishop Thomas Sprat, who had hitherto been staunch in the King's support, but was something of a weathercock for changing fortunes, now caved in. On 24 August, Sprat refused to sit on the Ecclesiastical Commission any more. His reasons were that he disapproved of the Commission punishing those clergy who had refused to read the King's Declaration of Indulgence in May and June; 'though I did myself submit in that Particular, yet I will never be in any way instrumental in punishing those [of] my brethren that did not'.[36] In a later defence of his actions during the reign of James, Theophilus Earl of Huntingdon also claimed to have resigned from the Commission at this time because he had become 'much dissatisfied' the King's policies.[37]

On 30 August, further medals were struck of the seven bishops, with their portraits on one side and Jesuits with pickaxe and shovel on the other trying to pull down a church. The medal was stamped 'The Gates of Hell shall not prevail against it'. Sancroft, hearing that the Catholic vicars-apostolic were still touring England, sent the bishops to their dioceses with instructions to confirm as quickly and rigorously as they could. This would also enable the bishops to canvass the views of leading laymen in their dioceses.[38] It was reported in London on 28 August that 'the Lord Bishops that were lately in the Tower have all of them redoubled their assiduity in preaching twice a day and going about confirming etc' and nine days later 'the Roman clergy about London in imitation of, as well as, ye Protestant Bishops begin their several circuits very speedily in order to confirm ye youth'.[39] The bishops remained heroes of the mob. A story recounted by Macaulay was that Bishop Cartwright, mistaken for one of the seven, was asked for his blessing. When the man who sought it discovered that it was Cartwright, he gave his blessing back.[40] Printed accounts of the trial of the bishops were enormously popular and sold well in the late summer of 1688.[41] This suggests that John Spurr was right to claim that 'the support for the seven bishops indicated that the Church was popular, perhaps more popular than it had been before'.[42]

Antagonism towards Catholics seemed to be the obverse side of the coin of popular support for the seven bishops. Late in September 1688, Father Charles Petre, brother of James's adviser and confessor, was preaching against the King James's version of the Bible when a mob pulled him from the pulpit and treated him 'coarsely'. A few days later, a Catholic house in Clerkenwell was attacked by the mob, in it gridirons, knives and caldrons were revealed which, it was suspected, would be used against Protestants. In Norwich, Newcastle, Cambridge, Bristol, Oxford and York, there were anti-Catholic riots in which priests were abused. Prudent tradesmen took down any signs that referred to Catholicism, such as 'the cardinal's hat', 'the nun's head' or 'the pope's head'.[43] One historian has argued that the government lost control of the country after the acquittal.[44]

The acquittal of the bishops also reverberated through the law courts for some time. On 11 July, the grand jury of the city of London was sent out three or four times to find a bill against those charged with riot in the wake of the acquittal, but would not do so. At the same session, one of the jurors in the case of the seven bishops was prevailed on to publicly recant his verdict and to champion the case for repeal of the Test Act, but it earned him no credit. The King also instructed the judges to declare that he had proceeded against the bishops only because they had been truly guilty and that he intended to proceed against them at the next meeting of Parliament. Sir Richard Allybone condemned the acquittal of the seven bishops in a charge at the Croydon assizes; 'if the King had been a Turk or Jew, it had all been one; for the subject ought to obey' Allybone said. It was claimed that Allybone's death a few days later was a result of 'overheating himself by his vehemence in declaiming against the Bishops'.[45] Judge Rotheram also claimed that the bishops' actions 'tended to bar the prerogative and alter the government which is treason'. Later, at Gloucester, Rotheram was insulted by Bishop Frampton of Gloucester, who preached on 'the consequences of stretching a conscience for any worldly preferment'.[46] In Oxford, Rotheram 'enveighed briskly against the Church of England and its clergy, who discover, he said, ye spirit of persecution'. However, he was also obliged to deal with trivial cases, such as two scholars indicted for insulting Obadiah Walker.[47] Judge Richard Heath, who went on circuit in August 1688, suggested in his charge to the juries that those lighting bonfires to celebrate the acquittal of the bishops should be indicted for riot. Judges repeatedly told grand juries that whatever the court might have decided the bishops were guilty.[48] In Hampshire, the grand jury stubbornly refused to bring indictments against those who celebrated the acquittal of the bishops, and refused to endorse any abhorrence of the bishops' petition.[49] As a consequence, there were those who questioned whether the judges could carry out James's wishes.[50] Havighurst claimed that, as a result of the trial of the bishops, 'the prestige of the judges was gone'.[51]

Despite his poor showing, William Williams, the Solicitor-General, was rewarded with a baronetcy while James, in revenge for their judgements on the side of the bishops, removed Judges Powell and Holloway from the judicial bench. Since Holloway and Powell became heroes themselves after their role in the trial, they often appeared in the memorials to James and to William.[52] Even before the trial, Powell had shown himself equivocal to the notion of a suspending power in ecclesiastical matters. One of the Verneys' correspondents wrote, 'Powell spoke so much that some asked if he were Advocate for the Bishops'.[53] On 6 November, when the bishops met the King, they complained that since their acquittal, they had been condemned and insulted by remarks of the judges during their assize tours.[54]

For the bishops' defence lawyers, things went differently. Sawyer and his co-counsel became the darlings of Protestant crowds. In September 1688, Sawyer planned to stand as MP for the University of Cambridge, suggesting to Sancroft that he did so to dishearten the 'enemies of our religion'.⁵⁵ Somers was equally popular. He was seen as a possible MP for Worcestershire; he had previously been on James's list of acceptable candidates for Parliament, but was quickly removed from it. At the restoration of the city of London's charter in October 1688, Somers was offered the recordership; in November 1688, he was elected recorder of Worcester, and then on 11 January 1689, chosen as MP for Worcester in the elections to the Convention.

As it had in the run-up to the trial, James's propaganda machine and his supporters continued to pour out attacks on the bishops. In August, a parson in Hereford diocese, where Bishop Croft had supported reading the Declaration, wrote *A Prophylactick from Disloyalty in these Perilous Times* on the text from Proverbs, 'the Wrath of a King is as messengers of Death, but a wise man will pacifie it'. The parson gave thanks to Bishop Croft for saving him from 'the great sin of disobedience to God's vicegerent', quoting the authority of St Paul's injunction to Christians to submit themselves to the authority of higher powers. He wished that his fellow clergy had paid as much obedience to King James as to 'a most scandalous usurper' – presumably Cromwell. He quoted Sherlock's writing on non-resistance, and argued that disobedience was the same as rebellion and cited the Ecclesiastical Commissioners as claiming that refusal to obey was 'a manifest contempt' of the King's authority. The seven bishops, claimed the parson, had been wilful in their rebellion and hence he took the step to 'follow my King' and reject the 'obnoxious judgments of all the other bishops of Christendom', even though it earned him 'obloquy and railing' from other clergy.⁵⁶

Much of James's propaganda continued to target the forlorn hope of achieving a breach between the Anglicans and Dissenters. In September 1688, the King authorised the publication of *Nahash Revived*, in which Anglicans were depicted as 'hardened in their impieties and have given the lie to their former pretences of loyalty, passive obedience and non resistance'. The author argued that the Church of England's moderation toward Dissenters was 'contrary to the experiences of all since the Reformation'. He wrote, 'happy were the nation if all our present prelates were of the same spirit of the prelates in those days' but the 'malignancy of the spirit of persecution' was still ruling the Church. In an ingenuous argument, the author claimed 'if His Majesty had designed the subversion of the Reformed, and the establishment of the Popish [Church], as the national religion, he had a much fairer opportunity after the defeat of Monmouth and Argyle than now he has'. When the King granted his 'clemency' to Dissenters, the Anglicans changed their tone, and now they behaved with 'impudence'. The Dissenters were in danger of 'injustice and ingratitude' if they did not embrace the King's Declaration.[57] This propaganda did not go unchallenged, as James's election agent in Somerset reported in September: 'the books that have been dispurst have had a very good effect, though great endeavours have been made by the Church party to disswade people from reading them'.[58]

The wooing of dissent

During the stormy months that followed the acquittal, Sancroft carefully avoided entanglements in politics, focusing instead on the maintenance of good order in the Church. He also planned to improve relations with the moderate Dissenters, by making concessions which would remove some of their objections to the liturgy of the Church of England. The Archbishop called together leading London clergy – Tenison, Patrick, Sharp and Wake, those who had most credit with Dissenters – and 'the chief of the Dissenting ministers' to, as William Wake said, 'agree such points of ceremonies as are indifferent between them, and to take their measures for what is to be proposed about religion at [the] next Parliament'.[59] There was talk of revising prayerbook offices and liturgical revision, including the eucharist; this was the principal issue on which the Dissenters and Anglicans had disagreed at the 1661 Savoy Conference. Among the clergy working on the revisions, were those who had established a reputation for cooperation and sympathy to Dissent: Claggett, Sherlock, Scott and Tenison. Wake recalled,

> I cannot but think that the Archbishop had some good reason for setting us at work at such a busy time, upon this scheme. And I well remember that asking another of my Lords the Bishops who had been sent to the Tower what he thought of the times, he told me we should soon see: and when I pressed for a reason, gave me only this turn, that men who rode over precipices would either in a little time break their necks, or they would come to their journeys end.[60]

This represented a major departure for Sancroft, who had previously been a firm adherent of the monopoly of Anglicanism, and had been willing to see Dissenters persecuted if they sought to challenge it by worshiping openly. Sancroft's pragmatic stand on cooperation with Dissent represented an important moment. Essentially Sancroft had come to the view – long held by Bishop Compton – that the Church shared an interest with Protestant Dissent to resist James's Catholic policies, which was preferable to surrender to the King. Comprehension was now a matter of Church policy to hold back the aggression of James's policies. Thus Sancroft was prepared to compromise his position on Dissent in his stand against Catholicism.

The bishops were quite open in the promises they made to the Dissenters. One leading Dissenter wrote,

> I do assure you and I am certain I have the best grounds in the whole world for my assurances, that the bishops will never stir one jot from their petition; but that they will, whenever that happy opportunity shall offer itself, let the Protestant Dissenters find that they will be better than their word.[61]

Gilbert Burnet made public promises that the spirit of persecution was over, and that there was a new spirit of friendship between Anglicans and Dissenters.[62] Burnet's words seemed to be born out by the marginalisation of the bishops most prone to persecute Dissenters such as Watson of St David's.

At the same time, James also redoubled attempts to woo the Protestant Dissenters. At the end of July, he had admitted some to the Privy Council; Evelyn noted that

> Colonel Titus, Sir Henry Vane (son of him who was executed for his treason) and some other of the Presbyterians and Independent party, were sworn of the Privy Council from hopes thereby to divert that party from going over to the bishops and the Church of England, which now they began to do: foreseeing the designes of the Papists.[63]

But the Dissenters remained convinced that James's concessions were only under 'threat from abroad' rather than a genuine commitment to toleration.[64] Across the country, evidence emerged of the combination of Anglicans and Dissenters against James. In Derbyshire, for example, in August 1688, John Gisborne wrote to Sir John Gell that if there was an election, they should seek a compromise with the Dissenters to avoid a contested election.[65] At least one publication suggested that few were taken in by James's strategy to woo Dissent. *Ten Seasonable Queries, proposed by a Protestant, that is for the Liberty of Conscience of all Perswasions* asked some pointed questions. Had not James supported the persecution of Dissent under Charles II? Was he using the Dissenters to 'pull down the Church'? Would he be for Liberty 'any longer than serves his turn?' Whether his 'prosecuting the seven bishops was... sufficient reason how well the King intends to repeal the Declaration for Liberty of Conscience?'

In August and September, James continued his campaign to pack Parliament by amending town charters and by forcing compliant candidates on constituencies. But the popular mood generated by the trial was looking towards William. A ballad to the tune of 'Couragio' urged

> Come, come great Orange, come away
> On thy August Voyagio
> The Church and State admit no stay
> And Protestants would once more say
> Couragio, Couragio, Couragio.[66]

Edmund Bohun noted that 'all men desired nothing more' than William's arrival and 'rejoiced at it as a deliverance sent by God'.[67] It seemed to some that time was running out; on 26 September, Bishop Lloyd of Norwich wrote to Sancroft saying that James's policies appeared to be working: in Norfolk, only the two Norwich seats would be likely to elect supporters of the Church of England.[68] As time went on, James's election agents were optimistic, but it was clear to others that James had little chance of obtaining a compliant Parliament. Sir John Baber, a Presbyterian with good connections at court, told the French envoy Bonrepaus that James would not be able to force the repeal of the Test Act through the Parliament. Sunderland also felt that this was the case.[69]

The Anglican revolution in the state

Thomas Ken tried to avoid being sucked further into the crisis by returning to his diocese. When the preparations of William of Orange for the invasion of England roused James, he made an attempt to conciliate the bishops. Sunderland was commanded to write to Ken, indicating that the King wished to confer with some of his bishops, and requested his attendance in London. Ken returned to London and, as Sancroft was ill (which he was frequently that autumn whether diplomatically or not) went with five other bishops to Whitehall, where they had audience with the King.[70] Unfortunately, James had changed his mind, and confined himself to reminding them of the duty and loyalty they owed to him; though he conceded that the issue of the petition was 'buried' as far as he was concerned.⁷¹ Ken expressed his disappointment 'that his Majesty should have required them to come so far in order to repeat to them what they so well understood before'. Sancroft waited on James the next day, to ask for another audience so that he and the other bishops might explain themselves.

At the same time, James issued a Declaration on 'Universal Liberty of Conscience for all Our Subjects'. In it James promised,

> That it was his purpose to endeavour a legal establishment of Liberty of Conscience for all his subjects, and inviolably to preserve the Church of England by a confirmation of the several acts of Uniformity. And for the farther securing not only the Church of England, but the Protestant Religion in general, he was willing the Roman Catholics should remain incapable of being members of the House of Commons, to remove those fears and apprehensions, lest the legislative power should be ingrossed by them, and turned against the Protestants, he farther assured his loving subjects, to do everything else for their safety and advantage that became a King.⁷²

Advised by Sunderland, James was in reverse gear and promised to consult the bishops. The Catholic Admiral Strickland was dismissed from the navy and replaced by Lord Dartmouth, and writs for the new Parliament were recalled. Astonished at the King's change of tack, Clarendon wondered what the bishops would do. When Bishop Francis Turner came to meet the King, he found that his Catholic supporters had regrouped and James was no longer quite as desperate to please as he had been a few days earlier.

On the same day that James issued his new 'Declaration on Universal Liberty of Conscience', an account was published in which Bishop Henry Compton and Anglican and Dissenting clergy of London were shown addressing William of Orange, offering their 'Humble Duties and Grateful Respects for his great and most hazardous undertakings for their deliverance...That they gave daily many thanksgivings to Almighty God...to preserve his person and favour his good design.'⁷³ But many felt unclear as to what might happen, John Gadbury wrote to Wales: 'we are still in ye dark in great measure as to ye Dutch fleet'.⁷⁴

In September, William issued a Declaration of his own, asserting that James had broken the laws of England, invaded the liberties of the Church by committing it to the clutches of the Ecclesiastical Commission and intruded Catholics into office. William indicated that he had been asked by lords spiritual and temporal to come to England to secure a free parliament and to safeguard the Church. Contained in the Declaration was that the Ecclesiastical Commission had illegally suspended the Bishop of London for the offence of refusing to obey an illegal order.[75] Although it was issued in London by Dutch agents, William brought hundreds of thousands for distribution when he landed at Torbay. William had also made arrangements to distribute 1,500 copies to the Royal Navy. In consequence, Evelyn found the court in 'utmost consternation'.[76]

Late in September, James had another audience with Bishop Francis Turner and again expressed his desire for reconciliation with the bishops, he said 'the Bishops and I have been old friends, and if such have differed they may easily be good friends again'. But Turner could not be so easily mollified; he told the King that many Anglicans feared he was preparing to appoint Catholics and Dissenters to Anglican benefices. Turner also pointed out that James had granted dispensations to a number of Anglican clergy who had converted to Rome. The incumbents of All Hallows in Lynn, Putney and a living in Essex had become Catholics and now held Anglican benefices under the King's dispensation. James promised to grant no such future dispensations.[77]

When James became aware of Sancroft's proposals for comprehension of Dissenters and William's Declaration, he summoned the Archbishop to Whitehall, 'accompanied by all the bishops who were in town, [to] give him their candid advice in the present emergency'. Consequently, on 3 October 1688, Sancroft, with the Bishops of London, Winchester, St Asaph, Ely, Chichester, Bristol, Bath and Wells and Peterborough, waited on the King. The meeting occurred, inauspiciously, just after James had asked the pope to act as godfather to his infant son. During the meeting, James ranted that William was 'worse than Cromwell'.[78] James had asked for the prayers of the Church of England and Ken read the draft that Sancroft had drawn up, 'beseeching God to give His holy angels charge over the King, to preserve his royal person in health and safety, to inspire him with wisdom and justice, and to fill his heart with a fatherly care of all his people'. He also agreed a prayer 'for peace, and the prevention of bloodshed in the land, for the reconciliation of all differences and dissensions, and for the preservation of our holy religion, our ancient laws and government, and for universal charity in the same holy worship and communion'.

The main purpose of the meeting could not avoid humiliating the King with the bishops' demands. Sancroft read a paper containing the articles of advice they asked James to adopt. It showed how far the bishops had come, it began

> May it please your Sacred Majesty,
>
> When I had lately the Honour to wait upon you, you were pleased briefly to acquaint me with what passed two days before, between your Majesty and those [of] my Reverend Brethren: By which, and by the account which they themselves gave me, I perceived that in truth there passed nothing but in very general terms and expressions of your Majesty's gracious and favourable inclinations to the Church of England, and of our reciprocal Duty and Loyalty to your Majesty: Both of which were sufficiently understood and declared before; and (as one of my Brethren then told you) would have been in the same state if the Bishops had not stirred one foot out of their Dioceses. Sir, I found it grieved my Lords the Bishops to have come so far, and to have done so little; and I am assured they came then prepared to have given your Majesty some more particular Instances of their Duty and Zeal for your service, had they not apprehended from some words which fell from your Majesty that you were not then at leisure to receive them. It was for this reason that I then besought your Majesty to command us once more to attend you altogether which your Majesty was pleased graciously to allow and encourage. We therefore are here now before you with all humility, to beg your permission that we may suggest to your Majesty such Advices as we think proper at this season and conducing to your service...

The claim of humility paled, however, when the bishops presented ten 'advices'. These were: only to allow those legally qualified under the Test Act to hold government places; the annulment of the Ecclesiastical Commission; the abandonment of the dispensing power and the restoration of the fellows of Magdalen College, Oxford; the revocation of licenses for Catholic schools; referral of the issue of the dispensing power to Parliament; prevention of the four vicars-apostolic from acting against the interests of the Church of England; filling the vacant Anglican livings and sees with some 'worthy persons'; abandonment of the *quo warranto* changes to charters and the restoration of boroughs' ancient charters; the calling of a free Parliament, and finally that he permit the bishops to present the case for his return to the Church of England.

Of these, three advices were immediately acted on: the Ecclesiastical Commission was dissolved, the Magdalen College fellows were restored and the Charter of London was restored.[79] The bishops were then asked by James to prepare a form of public prayer, to be read in all the churches, for averting the dangers which threatened the nation. The bishops' response was to pray for the preservation of religion.

If the bishops were equivocal, so was James, he told D'Adda to reassure the pope that he had not given up Catholic interests in England. But, after the meeting, Barrillon predicted that James would have to come to terms with the bishops.[80] Certainly Godolphin told Turner that the King was at bay, he said, 'whatever now was fit to be asked by us wee might have it at least granted by degrees'.[81] Tim Harris claimed 'it appeared that the bishops had brought James to heel. If their programme had been carried out in full, there would have been an Anglican revolution that would have effectively resulted in a return to the status quo ante'.[82] This is true, but Harris did not indicate whether he regarded James's commitments as sincere, or simply temporising. If James had genuinely abandoned his agenda, Harris was right, but to believe James had done so without evidence would be rash. James had already indicated he would restore Compton to episcopal functions and was willing to abandon the Ecclesiastical Commission. It was conceded, 'nothing contributed more toward the dissolution of the Ecclesiastical Commission than the acquittal of these prelates'.[83] The King hoped concessions would bring those who had abandoned him back, he sent Lord Dunmore to the Earl of Derby to tell him to come to London, where he would 'find very different measures taken towards those practised when my Lord Derby was last there, which was a great joy to them all, viz that the Bishop of London was restored, that Magdalen College would be, that the seven bishops were much satisfied'.[84] Lord Dunmore told Lord Derby 'things begin to change mightily, for the Church of England is wholly trusted'.[85]

The bishops, having won a victory against the King, took the opportunity to maintain their propaganda campaign to reassure the Dissenters, as they had throughout the events before and after the trial. Within a day or so of their visit to James, *An Account of the Late Proposals of the Archbishop of Canterbury with some other Bishops, to his Majesty in a Letter to M. B. Esq*, was published. It reported surprise at some of the 'ill constructions' that had been put on the fact that the bishops had attended on the King 'especially considering that most of them are the very men who not many months ago appeared so publicly and so courageously, even to the hazard of all the interests they had in this

World, in defence of our Protestant Religion and the Law of the land'. The *Account* then gave details of the meeting, including the ten articles the bishops had asked of James. Though it denied that information had come from the bishops themselves, the tract concluded:

> In the mean time let You and I commend the Prudence of these Excellence Bishops, admire their courage and celebrate their just Praises, and never forget to offer up most fervent Thanks to God, for his adorning the Church of England, at this juncture, with such Eminent Apostolical Bishops.

The account of the bishops' meeting with James was widely disseminated. There was an assumption that the bishops' trial had justified their right to make demands on the King. There were fears, especially among the Dissenters, that the meeting between the bishops and the King had restored an aggressive Anglican monopoly. But the *Account* included the reassurance: 'shew this letter to all your friends, that some may lay down their fears...let the Protestant Dissenters find that they will be better than their ward given in their famous petition'.[86]

The King was as good as his word in dissolving the Ecclesiastical Commission, restoring Compton, reinstating the fellows of Magdalen and appointing Thomas Lamplugh to York. At Oxford, all the Catholic fellows were evicted and their predecessors reinstated to the great rejoicing of the city.[87] But, in his maladroit way James, issuing a general pardon for criminals, appeared to except from it any clergyman convicted of an offence. For some days, the confusion of whether clergy were exceptions to the pardon soured the effect of the action.[88] What James could not see, but which John Evelyn could, was that his best course of action was to engage 'the late imprisoned bishops' to 'reconcile matters'.[89] Barrillion told Louis that James was confident that he had won over the bishops, but he also told the King that James's concession of a new Parliament was a stratagem rather than a genuine offer.[90]

When the Ecclesiastical Commission was dissolved in October 1688, Crewe and Cartwright were issued with general pardons to cover any wrongdoings they might be charged with during their time as commissioners.[91] On 17 October, James conceded the bishops' demand that he annul all the changes to borough charters imposed since 1679. Ejected lords lieutenants and magistrates were also restored to their office.[92] But, even now, some would not comply, in London Sir William Pritchard, who had been lord mayor under Charles II and was brought back under the reversal of the *quo warranto* changes to the charter of the City of London, refused to act as mayor, three other erstwhile mayors also refused to serve. Eventually, Sir George Treby accepted the mayoralty.[93]

James, however, was not entirely cowed by his agreement with the bishops. His memoranda to election agents suggested that he was still bent on circumventing a free election. He urged them to get to know those who were opposed to the Test and persuade them to stand for election. The King went on, 'you shall take care to make all persons understand that the late proceedings against the Bishops were necessary to support His Majesty's Declaration for Liberty of Conscience, which the King will always maintain, as likewise his prerogative on which it is founded'. He argued that the bishops' petition was designed to obstruct the meeting of Parliament 'which is so far from discouraging His Majesty that he is more resolved than ever to pursue this great work, not doubting to effect it, whatsoever opposition he may meet with'.[94] Moreover, James continued to use the dispensing power to the advantage of his Catholic friends. On 13 October, James issued to Sancroft an order to give a dispensation to his Catholic chaplain, Robert Hanbury, to hold two livings against the law regulating pluralities. This was quite normal, but to seek to dispense with the law, and to Sancroft, so soon after the trial and the 3 October meeting was foolhardy.[95]

At a further meeting with Sancroft on 16 October, during which James shared with the Archbishop some intelligence on William's impending invasion, Sancroft, to the King's intense frustration, said that he did not think the Prince had such a plan. In such a tense atmosphere, even the King's birthday on 14 October passed without the usual salute of guns from the Tower of London.[96] The King ordered horses and oxen to be secured so that no draught animals might be available for William wherever he might land.[97] He also ordered JPs to close down all coffee houses and public houses that stocked newspapers.[98] In the middle of October, James had a series of meetings with Dissenters, first the Presbyterians then the Independents and finally the Anabaptists. To each, he promised that he had not abandoned his Declaration of Liberty of Conscience and it remained part of his policy. The problem was that James was promising everything to everyone, unaware that the Dissenters and Anglicans knew exactly what he was saying to each other.[99] James was sharp enough to realise that he needed a scapegoat, however, and, on 27 October, dismissed Sunderland from both the lord presidency and from membership of the Privy Council. He was replaced by the Catholic sympathiser Lord Middleton. But James was unable to satisfy moderates like Nottingham, who still refused to attend the Privy Council meeting, despite James's request that he do so.[100]

Preparations for the invasion

In October, some politicians who were working to support the arrival of William; Lord Danby, Lord Dunblane and Sir Henry Goodricke, met in Yorkshire. It seems likely that Bishop Compton also joined them. The meeting had been arranged to put in hand the necessary measures to help William's invasion. It ensured that copies of William's Declaration, with additional comments made on 24 October, were printed and circulating as widely as possible, and certainly sufficiently so for James to receive a copy.[101] Time seemed to embolden James's opponents. At a Privy Council meeting on 29 October, it was reported that 'the Archbishop, the Marquis of Halifax, the Earls of Clarendon and Nottingham, refusing to sit at the Council table in their places amongst Papists and their boldly telling his Majesty that whatever was done whilst such [men] sat amongst them was unlawful'.[102]

On 1 November, at James's behest, Compton was called to a meeting with the King. James directly asked him whether he was one of the lords spiritual mentioned in William's Declaration. Compton, not very subtly, avoided a direct lie by saying 'I am confident that the rest of the bishops would as readily answer in the negative as myself'. James, however, appeared to be quiescent, saying 'I believe you are all innocent', but asked Compton to make a public pronouncement abhorring the invasion plans and William's Declaration. Compton asked if he could see the Declaration so he could read what he was denouncing, but James refused to show him a copy, so Compton demurred. In irritation, James said he would ask all the bishops to answer for themselves.

Before he met the bishops again, James had an audience with Nottingham, Halifax and Burlington, three moderate Tories who seemed uncommitted to either William or James. James asked them the same questions he planned to ask the bishops: were any of them involved in supporting William's expedition? The three said they had not. When James then asked them to sign an address he produced abhorring the invasion, they refused to do so, so James asked them to draw one up themselves.

The King asked Sancroft and the other prelates again to meet him on 6 November, the day after William had landed in Torbay. James showed them a passage in William's Declaration, stating that he had been invited over by several of the lords spiritual and temporal. 'I am fully satisfied of the innocence of my bishops', said James, 'yet I think it only proper to acquaint you with this statement'. He asked them whether they had invited William. Only Compton, Bishop of London, who *had* written to William inviting him to come, evasively observed, 'I have given his Majesty my answer yesterday'. It had, of course, been completely couched in prevaricating words. Sancroft said, 'they had not, [invited William] for they were men of peace, and would not mix themselves up with politics'. Sancroft also protested, truthfully, that 'the assertion as regarded himself was utterly false', for that he had never had any communication with the Prince of Orange, nor did he believe that any of his brother bishops had given William such an invitation. Sancroft could in good conscience recite his sacerdotal view of kingship: 'for my part', he said, 'I have but one King, him to whom my allegiance is naturally due, and which I have voluntarily renewed in oaths of homage and supremacy'. The other bishops were necessarily evasive however; Compton was because he had invited William to come, Lloyd and Trelawny because they supported the invasion, the others because they knew that they only owed James's change of heart to the threat of William's invasion.[103]

The King asked Sancroft and the other bishops for their paper expressing their abhorrence of the Prince of Orange and his designs. According to Thomas Sprat, Bishop of Rochester, the exchange was uncomfortable. The bishops replied to James:

'Sir we have brought no paper.'

James said: 'But I expected a paper from you: I take it, you promised me one. I look upon it to be absolutely necessary for my service.'

The bishops replied: 'We assure your majesty, scarce one in five hundred believes [the public statement of William] to be the Prince's true declaration.'

James exclaimed: 'What? Must I not be believed? Must my credit be called in question?'

Sancroft's reply cannot have been made without a sense of retribution: 'Truly Sir, we have lately some of us here...so severely smarted for meddling with matters of state and government, that it may well make us exceeding cautious how we do so any more'.

James, correctly, took this to be a reproach for their trial, and snapped back: 'I thought this had been quite forgotten!...This is the method I have proposed. I am your King! I am judge what is best for me! I will go my own way: I desire your assistance in it!'[104]

The bishops were not to be cowed, Sancroft even took the opportunity of complaining of an affront they had received during the trial from one of the judges, who had ridiculed them by criticising the petition, saying 'that they did not write true English, and it was fit they should be convicted by Dr. Busby of false grammar'.

'My lord', replied the King, 'this is querelle d'Allemand, a matter quite out of the way. I thought this had been all forgotten. For my part I am no lawyer. I am obliged to think what my judges do is according to law. But, if you will still complain on that account, I think I have reason to complain too. I am sure your counsel did not use me civilly'. The bishops concluded by saying that they were ready to serve James either in Parliament or with their prayers. That Sancroft was inclined to comply with James's request was suggested by a document among his papers solemnly denying the allegation of William of Orange's Declaration. But he was probably dissuaded by the Bishop of London from putting it forth. The unanimous refusal of Halifax, Nottingham and Burlington and the bishops to denounce William's Declaration suggests an element of collusion in how they responded to James.[105]

The interview on 6 November confirmed how much events were running away from James, but also from the bishops. Sancroft, White, Turner, Ken and Lake were clearly still loyal to James, although still smarting from the trial and imprisonment, and perhaps willing to see the threat from William bring James to a more compliant relationship with the Church. Their hope was to settle on the King's agreement in October, and resume an Anglican settlement in the State. Lloyd and Trelawny, with Compton, were, by November, clearly adopting a more radical solution that would see the dispossession of James of his throne. The unanimity of the bishops had barely lasted a few weeks after the trial. Sancroft and the loyalists were still willing to resist James in small matters, such as prayers and the denunciation of William's Declaration, as it served to keep James on the defensive, but fundamentally they remained conservative in outlook.

7
The Revolution

The invasion

During what turned out to be the last weeks of James's reign, intelligence of William's actions was both good and widespread. On 25 October 1688, Richard Burd wrote to his friend in Hampshire, Thomas Jevoise, telling him that he had dined at the Earl of Carlisle's house and heard the Duke of Norfolk telling all and sundry that the King had told the Privy Council that 'although the Prince of Orange was not yet come, yet that he went on shipboard last Thursday, and resolved the next easterly wind to be in England. Therefore the Dutch will most certainly be here the next fair wind without any further delays'.[1]

William chose his commanders astutely. His naval commander was Arthur Herbert and his general was the Frenchman Marshal Schomberg. Schomberg, who had been Louis XIV's commander, was exiled from France owing to his Protestantism. This had earned Schomberg a great following in England. James, on the other hand, found that his campaign to purge the counties of lords lieutenant as a prelude to creating a compliant Parliament had damaged his militia and local support when it was most needed. James had swept over 20 experienced lords lieutenant from power, and replaced them with men often without county connections and certainly without experience in mobilising and commanding the militia. Despite their restoration in October, during the emergency few militias were able, or willing, to respond to the King's summons. The Kent militia was unable to be called out at all. Lord Lindsey in Lincolnshire archly wrote to the King that, as the militia was 'not pleasing to your majesty', it had not been mustered since 1685. The Earl of Bristol similarly told the King that it would take three or four months at least to bring the militia up to strength and readiness.

When the winds on the North Sea and the Channel turned from West to North to East, William and James claimed, in turn, they were Protestant and Popish winds. James was perhaps more earnest in such claims, having ordered the sacrament to be carried in procession for three days, in hopes of changing the weather. Others took more practical measures; Lord Godolphin advised his sister to leave London as 'the Dutch will probably land here in England within less than a week'.[2] William had originally considered landing on the north-east coast of England, and Danby had gone there to prepare to raise the north in William's favour.[3] Compton, as bishop, aristocrat and a victim of James's extreme policies, was uniquely qualified to act as an intermediary and be trusted by many groups; Danby, Devonshire and Lumley all conspired with him as he toured the Midlands and North in late September 1688.[4]

The invitation to William had emphasised the navy's disaffection to James, and certainly Dartmouth, its new commander, found that some seamen refused to fire on Dutch ships. In the event, the navy permitted William to slip past to the south coast, to which he had changed his landing point. On the auspicious date of 5 November, William landed at Torbay. Thereafter, travelling through Devon in the most appalling weather, William found ready-made places of hospitality for him. At Forde Abbey, the Courtenay family greeted William; the elderly William Courtenay was a survivor of both the Rye House Plot against Charles II and the Monmouth Rebellion, and fiercely opposed James. At Chudleigh, he stayed with Mrs Gawler, widow of a Monmouth rebel. And, in his advance guard, Captain Hicks acted as a guide for the army, Hicks was the son of a Nonconformist minister executed at Taunton after the Monmouth rising.

When, on 9 November, William entered Exeter, the Bishop, Thomas Lamplugh, fled to London. William claimed that Lamplugh had deserted his flock, which, he asserted, was now committed to his care. This was a model for William's later approach to James's desertion of the country. William permitted the clergy of Exeter to pray for the King and the royal family, but ordered them to omit any prayers for the Prince of Wales.[5] The mayor of Exeter, who had been twice displaced and reinstated during James's charter revisions, refused to cooperate with William but the city put up no resistance to him. In Exeter, William had a banner made and inscribed with the words 'God and the Protestant Religion'.[6]

The rider who brought the news to James of William's invasion rode seven horses to death to get to London in just 30 hours. Within a day,

James called a council meeting to denounce the invasion as 'unchristian and unnatural' and a usurpation of his rights and authority. James also denounced William's claim that he came to establish a free Parliament and to defend the Church. But when James had a meeting with his former minister, Sunderland, the latter spoke to the King as if William's success was inevitable.[7] Bishop Lloyd went into hiding and advised Sancroft to find a place of safety.[8]

As news of the invasion spread, even those areas thought to be loyal to James, such as the north, remained quiet. Sir John Reresby in York recorded that no one seemed concerned about the landing, and most people said that the Prince only came to support the Church and intended no harm to England herself. Reresby was clearly in ignorance of Danby's advanced plans to raise Yorkshire and the north for William. Those of James's opponents who had not been let into the conspiracy feared that the King himself had orchestrated the invasion to give him a pretext to bring Irish troops to England and subdue the country by force. Much of what little conflict there was during the invasion reflected religious divisions. In Cheshire, a Catholic stronghold, Lord Delamere, who had been tried and acquitted in the wake of the Monmouth Rebellion, found his house attacked by Catholic forces before Protestant supporters came to his rescue. In a later speech, which was widely circulated, Delamere said, 'I am to chuse whether I will be a slave and a Papist or a Protestant and a freeman...If the K. prevails, farewell Liberty of Conscience'.[9] Attacks on Catholic chapels took place in both Oxford and Cambridge, where there was anti-Catholic unrest for some days, as well as in Newcastle, Uxbridge, Bristol and various towns around London. In most cases, chapels were pulled down and ransacked.[10] The relief some felt at William's invasion is best summed up by Sir John Bramston: 'had not the Prince come and these persons thus appeared, our religion had been rooted out'.[11]

When William landed at Torbay, Bishop Ken, who feared that he would soon be advancing on Wells, immediately left the town, and wrote to Sancroft, on 24 November, demonstrating even at this point that he was loyal and committed to James:

> I received intelligence that the Dutch were just coming to Wells; upon which I immediately left the town, and in obedience to his Majesty's general commands, took all my coach horses with me, and as many of my saddle horses as I well could, and took the shelter of a private village in Wiltshire, intending, if his Majesty had come into my country, to have waited on him and to have paid him my duty. But this

morning we are told that his Majesty is gone back to London, so that I only wait till the Dutch have passed my diocese, and then resolve to return thither again, that being my proper station. I would not have left the diocese in this juncture, but that the Dutch had seized houses within ten miles of Wells before I went; and your grace knows that I, having been a servant to the princess, and well acquainted with many of the Dutch, I could not have staid without giving some occasion of suspicion, which I thought it more advisable to avoid; resolving, by God's grace, to continue in a firm loyalty to the King, whom God direct and preserve in this time of danger; and I beseech your grace to lay my most humble duty at his Majesty's feet, and to acquaint him with the reason of my retiring, that I may not be misunderstood. God of His infinite mercy deliver us from the calamities which now threaten us, and from the sins which have occasioned them.[12]

In response to the invasion, James ordered his cavalry under Sir John Lanier and his infantry under Lord Faversham to muster at Salisbury. On 14 November, Colonel Kirk, commanding James's advance guard at Warminster, was also ordered to fall back to Salisbury.

The imploring of Catholics that James's presence was needed in London on 17 November – the anniversary of the accession of Elizabeth I to the throne – to avoid the usual anti-Catholic demonstrations on what was seen as a Protestant anniversary, delayed his journey to join his troops in Salisbury. James's advisers were right, London saw more than usual violence. There was also unrest in Norwich, Oxford, Northampton, Cambridge, York, Bristol, Bury St Edmunds, Gloucester, Hereford, Ipswich, Newcastle-on-Tyne, Shrewsbury, Stafford, Sudbury, Wolverhampton and Worcester. Harris regarded these disturbances as structured and disciplined with a directing intelligence. He argued that

> There was such widespread opposition to James's ambitious measures to help his co-religionists prior to William's invasion that there would have been some sort of revolution in the autumn of 1688 even if William had never set foot on English soil. Even William's 'conquest' of England can, in many respects, be regarded as being effected from below, since James was not overcome in battle by a foreign army but, rather, he fled in the face of widespread dissatisfaction amongst his subjects.[13]

In London, the printing house of Henry Mills, the King's Printer, which had poured out Catholic propaganda, was sacked by a mob of over a thousand. The unrest forced James to bring 5,000 troops to London to protect strategic points and the homes of leading Catholics and ambassadors.[14] Sir Edward Hales, the lieutenant of the Tower of London, even suggested to James that he mount the Tower mortars to face the City so that they could be used on the mob. James had sufficient sense not to follow such advice and, by the end of the month, Hales had been replaced by the more conciliatory Colonel Bevil Skelton.[15] James also received petitions from Nottingham, Halifax, Weymouth and Bishops White and Lloyd, seeking a promise to issue writs for a new Parliament. Irritated beyond endurance, James peremptorily refused.[16] Even Bishop Crewe, one of James's die-hard supporters, turned on the King and finally joined those who would no longer sit in the Privy Council with Catholics.[17] He had already issued a public petition to the King demanding his suppression of Catholic chapels, to appoint an archbishop of York and to call a free Parliament.[18] The last flickerings of good news reached London about this time; James heard, and recorded in the *London Gazette*, that Lord Lovelace had been arrested in his attempt to join William at Exeter and imprisoned by the Duke of Beaufort at Gloucester.

The rest of the news was unremittingly bad. Lord Cornbury, Clarendon's son, defected to William. He was joined, soon after, by Lord Abingdon and Sir Edward Seymour, these two were interesting defections. Abingdon had been lord lieutenant of Oxfordshire until he had been purged from office for refusing to support the repeal of the Test Acts. Seymour was a firm Tory, who had never supported William, but had been pushed to overcome his political principles and personal dislikes. Then came the defections of Lords Colchester, Wharton and Russell. Barillon wrote to Louis XIV that he had no faith that any of James's commanders would fight with their hearts for the King. News of two other defections came soon afterwards; Lord Delamere's was important because, as a great Cheshire landowner, he was able to close the Cheshire ports to the Irish troops that James hoped might come to his rescue. Second, the Duke of Beaufort, the restored lord lieutenant of Herefordshire, Gloucestershire and all of Wales, who had raised his counties for the King against Monmouth, was disheartened by the lack of support for James. He gave up his defence of the King in Bristol, and allowed Lord Shrewsbury to claim the city for William. Beaufort wrote to Middleton, the Secretary of State, of 'universal dissatisfaction of the next county [Somerset] and the lukewarmness of the best people here'.[19] Faced with the entire collapse of royal authority in the West, Lord Bath joined William and handed him control of the ports of Plymouth and

Falmouth. There were a few who stuck blindly to the Tory Anglican line, though Charles Hatton wrote to his father of James, 'knowing how firmly I have imbibed the principles of the Church of England, you will be secure I can never depart from my allegiance to my Prince'.[20]

James at bay

On 17 November, Clarendon presented the King with another address asking for Parliament to be called. But James angrily replied that he had no intention of calling elections while there were foreign troops on his soil. Within two days, events had pushed James to change his mind; he issued a 'Gracious Answer' promising 'upon the faith of a King' that he would call a Parliament 'as soon as ever the Prince of Orange has quitted this realm'.[21] He rode through snow and blizzards to get to Salisbury on 19 November and, despite the warm greeting of the people of the city, James was dejected. At a meeting with his commanders, including Churchill, Charles Trelawny and the Duke of Grafton, James told them of his change of heart and his new willingness to call a Parliament to consult with his subjects. But, in his depression, James's resolution seemed to crumble. A parson in Salisbury, Mr Chetwood, had been engaged to be James's Anglican chaplain, but he complained loudly when he found the chapel in the Bishop's Palace – where James resided – had been taken over by Catholic priests. Chetwood threatened to resign, and James was forced to remove his Catholic priests. When this was known, Chetwood was acclaimed as a hero.

James was plagued by nosebleeds and depression, and begged Barillon to ask Louis for help. Undecided as to whether to return to London or to remain at Salisbury, he ordered his infantry to return to London on 22 November. Later that night, Churchill and Grafton defected to William, the former leaving a letter saying that he was only motivated by principle not advantage; and Colonel Kirk at Warminster was refusing to obey orders. James, beset with a crumbling army command returned to London. His son-in-law, Prince George of Denmark, promised to follow but defected by turning north to join his wife in Nottingham. George, often written off as a nonentity, left James a letter blaming 'the restless spirit of the enemies of reformed religion, backed by the cruel zeal and the prevailing power of France' for undermining 'the laws and establishment of that government on which alone depends the well being of your Majesty and of the Protestant religion in Europe'. In London, James found that his daughter Anne had also flown. She had written to Exeter wishing William well and, together with Bishop

Compton in jackboots and a sword, left London for Nottingham on 25 November. One account had it that Compton's aid in the rescue of Anne 'helped him to a reverential sort of popularity which he of all ye bishops would least have found otherwise'.[22] Princess Anne also contributed to the events by publishing *Princess Anne of Denmark's Letter to the Queen*, a piece of shameless anti-James propaganda, which had been prepared the month before. In it, she affected surprise at her husband's actions but said that she stayed away from the King for fear of his displeasure. She also added her view that

> the general falling off of the nobility and gentry...have no other end than to prevail with the King to secure their religion, which they saw so much in danger,...I am fully persuaded that the Prince of Orange designs the King's safety and preservation...God grant a happy end to these troubles.[23]

On 22 November, Danby, at the head of a large body of gentlemen, rode into York, proclaiming 'A Free Parliament, the Protestant Religion and No Popery!'[24] Within a few days, Hull also declared itself for William. Danby's letter to the governor of Hull, Sir John Hanmer, illustrated the concerns of an Anglican Tory. He wrote

> were there any visible hopes under Heaven of saving the Protestant religion in England but by this opportunity that God has given us, I think you know me well enough to believe I am the last man in the kingdom that would attempt to have it rescued by force, but it has been made so plain to us what we are to expect from a power placed in the papists' hands that, whatsoever any man pretend, no man can in his heart believe that any man shall be able to find protection that will persist in the Protestant religion...I have reason to hope upon strong grounds...that there will not be a blow struck by the armies, but that the matter will be decided by Parliament.[25]

At the same time, Devonshire had secured Nottingham for William, and Lord Lumley secured Newcastle.

Before he left Exeter, on badly flooded roads through Honiton, Axminster, Hinton St George, Crewkerne, Sherborne to Abingdon, William issued a *Modest Vindication of the Petition of the Lords...for the Calling of a Free Parliament*, which declared that subjects had the right to petition the King and referred to 'prelates and patriots'.[26] While at Hinton St George, William received Dr Finch, Warden of All Souls

College, Oxford, who declared that the University was strongly in his support. At Berwick St James in Wiltshire, William was joined at last by Clarendon, who told Churchill of the rumours circulating in London that – before his defection – Churchill had planned to kidnap James and hand him over to William. On 26 November, Sunderland had his last meeting with James, who by now knew of Sunderland's crumbling loyalty, the King ruefully told Sunderland that he hoped he would be more faithful to his next master.[27]

On 27 November, James held a meeting with his remaining loyal peers and bishops. Tactfully, he omitted his Catholic councillors from the meeting. The other councillors 'opened their grievances boldly'.[28] Rochester demanded that James call Parliament and open negotiations with William. Clarendon – before he abandoned James – vented his spleen on James's Catholic policies, and also explained that, with the King wavering and indecisive, it was no wonder that so many were going over to William. Lord Ailesbury said it was like hearing a teacher speaking to a pupil. Halifax went even further, asking for all Catholics to be dismissed and for James to abandon links with France. James made the fatal error of saying that he was caught between choosing to retreat or to go abroad, before seeking his restoration 'as his brother had done'. Talk of fleeing abroad disheartened all present, but most people knew how much James was attracted to the image of his father as a martyr and his brother as a hero. He also muttered darkly about having read what happened to Richard II, meaning that he knew what happened to deposed kings who were imprisoned. After the meeting, the Imperial ambassador reported that James could remain King only on the basis that he would retain a shadow of his powers.

Soon after, James issued a Declaration which, 'considering the grievances and distempers of the people and the present circumstances of his Majesty's condition', conceded that Parliament would be called for 15 January. He also vainly let it be known to his former loyalists that he would not reek revenge on them.[29] At the same time – despite their fierce criticisms of the King – James assembled a commission of three, Lords Halifax, Godolphin and Nottingham, to meet William at Hungerford. Of the three, Godolphin was undoubtedly in contact with William through Lord Churchill.[30] James told Barillon that the talks were just a 'feint' that he had no intention of calling Parliament, and he was planning to leave England. He also said that any concessions he made were to buy time so that he could get the Queen and Prince of Wales to France. It was the sort of comment that undermines attempts to revise historical interpretations of James's character and motivation.

William, whose supporters felt that nothing was to be gained from such talks, delayed for some time. But, on 4 December, William entered Salisbury, where he was greeted with the same enthusiasm James had received. The next day, he set out for Andover to meet James's commissioners. On the way, he stayed with Lady Pye at Collingbourne, where the chaplain told a Dutch officer that it was a sin to oppose the sovereign. Two days later, Sir Charles Kemys was told by his servant, John Romsey, that Worcester was 'as loyal a town as any in England, *every man for the Prince*'.[31]

When the three commissioners met William, on 8 December, they brought two letters, one from James, which was conciliatory asking that obstacles to an agreement could be overcome and he signed himself 'votre affectionne pere et oncle Jaques R'. The second letter was from the three commissioners, stating that James had conceded the issue on which William had entered the country, namely calling a free Parliament, and the King was keen to resolve the issues before them. The posturing and positioning of the Prince and the commissioners was relatively pointless, no one really believed that William and James would come to a resolution of matters so easily. Aside from the main discussion, the most important conversation at Hungerford was between Halifax and Gilbert Burnet. Halifax asked Burnet whether the Prince wished to have the King in his hands. Burnet said William meant no harm to the King personally. Halifax asked 'what if the King had a mind to go away'? Burnet said nothing was to be wished more than for James to leave the country.

During the discussions, William moved his headquarters to Littlecote in Oxfordshire, the home of the Popham family. The Pophams had suffered during the Monmouth Rebellion, and had no love of James. Before long, Lords Abingdon and Clarendon joined the talks, on William's side, and there were open discussions about James's departure. Finally, on 10 December, William sent the three commissioners back to James with his reply. The demands William made may have been purposefully excessive, to push James into a rejection of them. William accepted the calling of Parliament, but required the dismissal of all Catholics from office, the recall of all proclamations against the Prince, William's supporters in prison were to be freed, the dockyards were to be placed in the hands of the City of London, and Portsmouth in the hands of William's commanders, William's army was to be paid by the Treasury, both William's and James's armies were to be kept 30 miles from London and William and James jointly were to preside at the first session of the new Parliament.

James did not take long to consider the terms William had laid down. By the time the three commissioners returned, the Queen and Prince of Wales had been spirited to France. On the day after the commissioners returned, James made arrangements to follow them. In self-pity, he asked Lord Ailesbury what he should do: his daughters, his army, his navy and the nobility had abandoned him. Ailesbury knew that James was planning to flee, and told him that he had heard of his secret plans to go. James asked Charles Bertie, Lord Danby's brother-in-law, whether he thought the northern nobles would come over to him and afford him some protection. Bertie's disheartening reply was that he did not think the northern nobles would actually harm him. Blaming the disaffection of his army and navy, James was determined to go.

Anti-Catholic disturbances continued across the country. On 11 December, the *Universal Intelligence* reported further unrest in York, Bristol, Gloucester, Worcester, Shrewsbury, Stafford, Wolverhampton, Birmingham, Cambridge, Hull, Newcastle, Northampton and Bury St Edmunds. At Clevely Park near Cambridge, Lord Dover's house, the chapel was burnt down. Cambridge was also the scene of the humiliation of Thomas Watson, Bishop of St David's. He was notorious for his support for James, and had prosecuted clergy for failure to read the Declaration of Indulgence. Watson was seized at Borough Green, mounted on a paltry horse, brought to Cambridge, where he was paraded through the streets and only rescued by the magistrates when they could secure him in the Castle. Similarly, rough treatment was meted out to James Arderne, dean of Chester, who had also supported James. Elsewhere, disobedience was targeted at Catholics for gain: in Maidstone five grenadiers and the youth of the town robbed Catholics. In Hull, the governor, the Catholic Lord Langdale, was ejected from the town without any ceremony.[32]

James fled London but was captured in a boat near Rochester and brought to a pub in Faversham. The mayor of Faversham, who had been imprisoned during the Monmouth Rebellion, kept the mob from him, but it took Lord Winchelsea, the lord lieutenant of Kent, to come to his rescue. Even Winchelsea, however, could not save James from the humiliation of having the gentlemen of East Kent reading out one of William's Declarations in front of him. Nevertheless, the rumours that the King had fled encouraged an outbreak of violence against the Catholics and ambassadors in London. Alone, the Spanish ambassador's losses allegedly amounted to £100,000. The Papal Nuncio and the ambassadors of Spain and Savoy fled the country, under passports issued by William.[33]

The Guildhall meeting

Following James's departure from London, the government of the country seemed to be vacant, and there were fears of mob rule in London. 'The rabble were the masters', as one account put it.[34] The void was filled by the bishops. Lord Rochester and Bishop Turner went to Lambeth, where Sancroft wrote to as many peers as he could, as 'Consiliarii Nati', to meet at the Guildhall; 'wee had otherwise bin in a state of Banditi, and London had certainly bin the spoyle of the rabble'.[35] James was later to say of this meeting, 'you were all kings when I left London'.[36] Turner defended himself claiming that there had been many instances in English history of peers assuming the government in the sovereign's absence. Nor, claimed Turner, did James disagree with anything done by the Guildhall meeting.

On the following day, 11 December, Sancroft, Lamplugh the new Archbishop of York and five other bishops, led a meeting of peers at the Guildhall in London, at which they assumed the government of the kingdom. They immediately declared that they had secured the government for the protection of the Church and to secure liberty of conscience for the Protestant Dissenters.[37] They issued orders to Lord Craven, lord lieutenant of Middlesex, to deploy the militia, and to Lords Feversham and Dartmouth who were told to stop military and naval actions against William. Lord Lucas was appointed lieutenant of the Tower. Robert Beddard's careful analysis of the Guildhall meeting shows that it was rife with tensions. The bishops, particularly Sancroft and what Beddard termed the 'Cavalier bishops', were especially confused. In 1679, they had opposed the Whig exclusionists on the grounds of James's legitimate succession; now at the Guildhall meeting they assumed the powers of the King, to whom they were determined to remain loyal, but whom they had so recently defied. They had been taken aback by the King's flight, but could not allow the anarchy that threatened the capital to take hold. If there was one thing that held the Tory and Cavalier bishops together ('the onely thing designed by many of us', as Francis Turner put it) it was to preserve the kingdom for James. Sancroft even came out of his self-imposed psychosomatic purdah for the meeting and took the chair. The Declaration of the Guildhall meeting was carefully constructed, expressing concern for the Church and for its liberties and resolving to assist William in establishing a free Parliament which would secure a settlement of laws, liberties, the Church and Dissent. It also announced the disarming of Catholics in London and Westminster, and ordered the fleet not to attack the Dutch. But it said nothing of the throne or James.

For Sancroft and the Tories, this was because they intended it would pave the way for James's return on their terms.

Numerically, the Tories could count on a majority at the Guildhall for James. Beddard claimed that, of 29 peers, 18 were Tories who were well-disposed toward James – or at least to the survival of his reign. Yet these loyalists were, like the bishops, divided and uncertain what to do. In contrast, the opponents of the King were united and determined that James should not return. Turner later saw this meeting as a crucial error. Rather than the Tories publicly and clearly asserting their desire for James to return, they allowed themselves to be sidetracked into other issues. When a bishop suggested the inclusion of a statement in James's interest, a lord loudly spoke of the 'bishops returning to their own vomit of popery'. This sort of open aggression to James from the Whig neo-exclusionists stopped the Tories in their tracks. Sancroft lacked the strength of leadership, and Tories, like Clarendon and Rochester, the will to confront and resist such views. Indeed, it seemed that the Tories had no defence of James to advance. Consequently, despite their numbers, the Whigs permitted no expression of loyalty to James, or of any wish for his return. When, at the conclusion, although Sancroft felt the meeting had not created the sort of Declaration he had wanted, he reluctantly signed it, saying that, if he did not sign it, he would be 'marked out'.[38] For a man who had so recently acted with such courage, his pusillanimity was extraordinary. Later, there were rumours that some at the Guildhall had been willing to 'relent for the King', but rejection of his Catholicism was unanimous.[39] The Guildhall meeting reflected a low point for the Tory Cavaliers and bishops, they lost the initiative and the Whigs grasped it and thereafter directed events.

Two days later, Ailesbury was woken with the news that the King had been captured. A troop of lifeguards was despatched to Faversham to bring James back to London. Perhaps in relief that the violence and anarchy was over, James was welcomed back to London by the population; though Edmund Bohun reported the people were cold and frowning when he rode through the city.[40] Coinciding with the King's flight and return, was growing anxiety at the rumours that Irish troops were coming to James's aid. Two or three times, Londoners had risen in the fear that Irish troops were already in or near the city; there were similar rumours across the country. Preston, Warrington, Wigan, Chesterfield and Leeds all saw unrest at rumours of Irish troops threatening the citizens. The effect of the unrest was to make people keener for the restoration of some order and control. While the Guildhall meeting had done so temporarily, William was gradually assuming such a role. From Henley, he began to issues orders as if he were the ruler. He ordered that disbanded troops from Feversham's royal army were to be rounded up and arrangements were made to protect the Queen dowager, Charles II's widow. Before William's troops entered London, it was said that he deliberately circulated rumours that Irish Catholic soldiers were on their way, to ensure that his Protestant troops were well-received.[41]

James's return presented William with a difficulty. There were various opinions on what to do with him: Lords Delamere and Mordaunt advocated throwing him into the Tower, but Grafton and Shrewsbury opposed this. William studiously refused to be drawn on what was best to be done with James. Finally, it was determined that James should be brought to Ham House in Richmond, but, in deference to James's own wishes, he was allowed to go to Rochester instead. William ordered that James should go at night, and by river, to prevent any public expressions of support for him; and James saw his own troops replaced by William's Dutch guards. When James left London, William installed himself at Whitehall Palace after a triumphal entry to the city.

After the Guildhall meeting, Sancroft appeared to undergo a collapse in confidence, and again withdrew from his public role. On 17 December, Bishop Lloyd was sent by a meeting of the others bishops to go to William and arrange a meeting between them. At that time, Sancroft had agreed to attend the meeting, but the next day he chose not to attend.[42]

On 21 December 1688, William of Orange issued a proclamation which referred to James's 'insufferable oppressions' of the Church of England, which had put his subjects 'under great fears'. James, claimed William, had acted contrary to law and, in referring to the seven bishops' petition, said that they 'were brought to trial, as if they had been guilty of some enormous crime'. The seven bishops, said William, had been forced to defend themselves before judges who were not qualified under the Test Act and 'by consequence were men whose interest led them to condemn them'.[43] This proclamation had been issued only four days after James had met the bishops and agreed the form and content of a different Declaration. In it, James agreed to respect the privileges of the Church, to confirm the Protestant monopoly on office, to grant control of ecclesiastical patronage to William and consent to all bills passed by Parliament. In short, James conceded all that the Church wanted. But it was too late for him to be trusted, even if he retained the personal authority to be obeyed.[44]

In Rochester, James increasingly felt the pressure of William for him to go. William permitted James to receive a letter from the Queen begging him to follow her; and, when James asked for passports, he was sent twice the number he sought. He was also advised by friends that he was not safe. Finally, William withdrew guards from the rear of the house so that James could slip unobserved onto the Thames. James fled on 23 December, leaving a letter explaining that he was only going because he knew his life was in danger.

On Christmas day, William pointedly attended the Chapel Royal at St James's Palace, where Burnet preached and Compton gave the Prince holy communion according to the Book of Common Prayer. For William, public receipt of the communion of the Church of England was an important signal. But prayers for the King and Queen were omitted. Parson Isaac Archer recorded in his diary, in December 1688, that he felt the country had been 'in danger of slavery and popery, if God had not sent the Prince of Orange on November 5th, who, without any bloodshed, hath, in some measure settled us. The King is gone and a convention is to be on January 22'. When William ascended the throne, Archer called it the dew of Heaven.[45]

James's flight heightened the issue of what should happen to the crown. Already, Henry Pollexfen, one of the bishops' lawyers, had advocated that there should be a grant of the crown to William. Bishop Lloyd was the only one of the seven bishops to agree wholeheartedly to the removal of James, and was the first to claim that James had abdicated.[46] Before James's flight, Bishop Turner told Lloyd that he believed that James had conceded all that William had come to establish. James had granted that he would call a free Parliament; he had pledged to remove Catholics from office and to protect the Church. It seemed likely that there was some support for Turner's views; the Earls of Clarendon and Rochester both felt that James might have learnt his lesson. However, the King's flight pulled the carpet out from under their feet. When Turner, who seems to have known James was planning to flee, asked what the bishops should do, James replied that they 'ought still to apply ourselves to ye Prince'.[47] As Louis XIV commented when he heard that James had landed in France: 'Le Roy d'Angleterre est un homme perdu.' It was not long before James's sister, the Duchesse d'Orleans, said that the more she saw of him the more excuses she found for the actions of William of Orange.

The convention

Above all, James's desertion meant that the initiative fell to those who remained in England. At four o'clock on Christmas Day, the peers in

London went to William and formally thanked him for coming to save the Church and establish a free Parliament. An election was the obvious solution, to allow a new Parliament to decide what should happen next. The problem was, of course, that James's *quo warranto* proceedings had muddled the franchise of boroughs so much that it was not clear how an election might easily be held. Consequently, when the peers also asked the Prince to move writs for a new Parliament, William made arrangements to convene all the MPs who had served in Charles II's time and he put the peers' request to them. This makeshift Commons agreed that a Convention be called for 22 January. At one of the meetings of peers, the Earl of Ailesbury – one of the few loyalists for James – looked around the room and decided that probably only he, and Sancroft, had not privately been in contact with William.[48]

Very quickly, four views emerged among those elected to the Convention. A small number of loyalists, including Sancroft, sought James's recall under his own pledges and undertakings; others, like Nottingham, wanted his return under stringent terms and conditions; some Tories wanted James to be excluded from the regnant powers, with a regent to act for him since he was incapable of exercising royal authority; and the final group wanted the throne to be declared vacant and William or Mary to be offered the crown. There were many variants of this latter view: Danby, for example, wanted Mary to be the monarch with William as her consort.[49] After Halifax declared himself for Mary to succeed, Burnet worked hard to spread the news that Mary herself had said she wished to defer to her husband as King. William's esteem for Gilbert Burnet rose when he pointed out that Phillip of Spain, Mary Tudor's husband, had enjoyed the crown matrimonial of England – and had been titled King. Burnet had ensured that Mary offered her husband the 'real authority' if she succeeded to the throne. The bishops, led by Sancroft, and with only a few exceptions, adopted the view that James should be recalled, with or without conditions. During the Convention, some members moved to the position of accepting the idea of a regency, but only as a rear-guard action to prevent the vacating of the throne.[50]

The calling of the Convention resulted in the publication of a number of letters and tracts. *The Letter to a Friend, advising in this Extraordinary Juncture how to Free the Nation from Slavery for ever*, included the question 'is the government dissolved, or only under some disorder?'[51] In *A Word to the Wise for Settling the Government*, the question was posed 'Whether [the] King hath not thereby made himself incompetent and uncapable to Govern a Protestant Church and a Protestant People by their Protestant Laws? And notoriously abdicated or renounced the Government?'[52] Most significant was *A Short Historical Account touching the Succession of the Crown*, which identified the elective nature of the English crown. Starting with Edward the Confessor, William I, William Rufus through most of the medieval kings to Charles II's restoration, the account indicated that there had always been a tradition of electing kings in England.[53]

On 21 January, the Convention assembled to debate the key question: was the throne vacant? The majority of the Commons was clear that James had abdicated; he had destroyed the writs for a new Parliament and by taking the matrix of the great seal with him, he had intended to leave anarchy in his wake. In such circumstances, James was held to have abandoned the government. The toughest resistance to this was from the Lords. Some, like the Earl of Pembroke, claimed that it was natural for James to flee to save himself. Finally, on 28 January, there was a vote on the motion 'that it hath been found by experience inconsistent with the safety and welfare of this Protestant kingdom to be governed by a Popish prince'.[54] While the motion that James had abdicated easily passed the Commons, in the Lords it squeaked home with a majority of three, with many of James's erstwhile supporters staying away from the vote. The bishops' position had been problematic. Turner conceded that those bishops who, like him, accepted the idea of a regency, only did so to gain time, and did so 'upon assurance enough that it would not be accepted'. So they grasped at any measure other than 'setting his [James's] crowne upon another head which except by force is never to be reverst'.[55]

In the joint meeting of Lords and Commons to manage the issue of the throne, Turner played a leading role. He claimed the King could legally leave the country 'if there be a going away with the purpose of seeking to recover what is for the present lost and forsaken'. Turner also argued that Catholicism did not, like lunacy and infancy, disqualify James from ruling; and warned against making the English throne elective, like that of Poland.[56] Lake was one of the bishops who advocated a regency on behalf of James's son, though the lingering questions about the legitimacy of the 'bedpan baby' made this an idea that few supported. In fact the idea of a regency was soon abandoned, with even Halifax and Danby conceding the Commons' view that a regency was unacceptable. Bishop Compton also opposed the regency, and Trelawny spoke against it, though he abstained in the vote.[57] In the end, the Convention declared that James had abdicated the throne. What none of the participants in the Convention openly expressed was the view that people were fearful that James's return would place them into the hands of a king who was untrustworthy and vengeful, qualities of which James had given them many examples.

News of the vote to vacate the throne quickly spread through the country. In Gloucester, the stone statue of James – erected only two years earlier – was pulled down and thrown into the Severn. But, among churchmen, there was no universal rejoicing. John Sharp, whose anti-Catholic preaching had been the cause of Compton's suspension, preached before the Commons, prayed for James and told the MPs to their faces that the deposing of kings was a popish and wicked action. The MPs were furious, and threatened to imprison Sharp. In the end, he suffered the snub of not having a vote of thanks for his sermon or a request that it be published. The next day, Burnet corrected things by preaching on the text 'Happy is the people that is in such a case: Yea happy is that people whose God is the Lord'. Burnet fully endorsed the eviction of James, and was thanked by a vote of the MPs.

As the Convention was meeting, a publication was produced on the 30 March 1689, which purported to be a letter from a bishop, and was the fullest expression of the High Anglican criticism of the pragmatism of the seven bishops. It raised the issue of the role of the Church and bishops in the political events of the time:

> Peradventure it were but justice to attribute mine, as well as the silence of divers of my brethren, to a prudence allowable in such an occasion. But some perhaps would think otherwise, and not without reason blame a man, who all his life has preacht obedience to soveraigns, and been brought up at Oxford, under men equally famous for learning and piety, if I did not follow their example. For in the time of a rebellion worse than this, they did not only testimony to the truth, when they were requested, but had the courage to defend it before the Usurpers of the Royal authority; and reproach their crime with a zeal worthy of our profession ... now we see bishops suffer themselves to bee drawn into the ways of the ungodly and sit in the seat of the scornful, I mean [those] who have joined with that wretched Convention.

The bishop asserted that passive obedience had always been the law of the Church, and every bishop has sworn to it in their oaths of allegiance. To participate in the Convention was to adopt the same mental reservation that had been the hallmark of Catholicism:

> Equivocation then, and mentall reservations are no longer to bee layed at the dore of the Papists, but of the zealots of the Convention,

who can no way, wash off the stain of perjury, but by declaring that they swore occasionally, that in engaged to be faithful no longer, than the King should be faithful to them, by performing an imaginary contract betwixt the King and the people.

The main thrust of the letter was that Protestant and Papist agreed that the only case in which it was not only lawful not to obey a ruler, but unlawful to obey, is when they order something contrary to the law of God. This principle led Sancroft and the six bishops to refuse to publish the King's Declaration. But the anonymous bishop held that, 'since the Church of England lodges the supreme power of governing this Church in the King, subject to none but God, I do not see that the right is altogether so cleer on their side as some pretend'. The question was whether to publish a declaration of their Prince by which the law was suspended 'which is a matter purely temporall, and in which by consequence the Bishops have no more authority than the temporal lords, that is, to propose in Parliament what they judge most for the public good'. The author conceded that the bishops had 'done the part of Good Pastors by hazarding themselves for their flocks, a glory which some flatter I fear more pleasingly than sincerely'. Their consciences might justify their intention but 'yet mee thinks that good conscience of theirs might have obliged them to explain themselves cleerly when they were examin'd before the Council, or try'd at Westminster-hall for a seditious libel'. Had the petition mentioned the original contract between the King and his people or a range of other issues 'do you, My Lord, believe they would have been so easily acquitted by the jury? Certainly not'. As the law stood, claimed the bishop,

> they would have been condemned as Traitors, and no body in a condition to help them... Let not those men, then, banter us longer with their pretended zeal, but think of asking God, and the King, pardon for their most detestable perfidiousness which ever blackened the reputation of our clergy.[58]

Of course the vote that the throne was vacant was not the same as granting the throne to William, since Mary was technically James's heir. William spent almost a fortnight waiting for Lords and Commons to decide what to do with the throne. In the end, he summoned Danby, Halifax and Shrewsbury – men he knew were undecided about the form of monarchy to be conferred on William and Mary – and told them he had not come to establish a commonwealth, or to be elected 'Duke of Venice'; he told them he would not accept either appointment as regent, or his wife succeeding on her own. He sought the award of the crown to him for life, or he would return to Holland and would leave the throne vacant. Halifax suspected the ultimatum was a bluff, but in reality there was no alternative. Danby threw his lot in with William.[59] Churchill managed to get Princess Anne to confirm that she would not accept the throne before William or Mary, though she sought succession for her children after William and Mary. Eventually, on 6 February, Parliament voted that William and Mary were to be joint sovereigns. In the Lords, the vote was 74 for, 38 against, and in the Commons 251 for and 183 against. Of the seven bishops, only Lloyd and Trelawny voted for William to succeed. Clarendon wrote that Lloyd 'told me that I was free from my oaths to King James, that he could very well take the new oaths, and that as things were, he took himself to be quite free from any obligation. Strange doctrine, as I thought, from a bishop'.[60]

Sancroft's departure

The day on which William and Mary were proclaimed King and Queen, Mary sent her chaplain, Dr Stanley, to Lambeth Palace, to ask for the archbishop's blessing for her. 'Tell the princess', replied Sancroft, 'to ask her father's; without that I doubt mine would be heard in Heaven'. Stanley had another errand to perform: attending a service in the chapel at Lambeth Palace to see whether prayers were offered there for King James or for the newly proclaimed sovereigns. Henry Wharton, Sancroft's chaplain, understanding Stanley's mission, asked Sancroft for his instructions. 'I have no new instructions to give', replied Sancroft, meaning that no alterations were to be made. But Wharton, who had resolved on taking the oaths of allegiance to William and Mary, and perhaps thinking he was helping Sancroft, prayed for William and Mary. Subsequently, Sancroft told him angrily that 'he must either desist from praying for William and Mary or cease to officiate in the chapel; for as long as King James was alive, no other persons could be sovereign of the country'.[61]

William was anxious to conciliate Sancroft, and nominated him one of his privy councillors, but he refused the office. Sancroft was begged by Danby and other members of the government to officiate at the coronation of William and Mary, but he refused either to crown them, or take the oaths required. 'How can he, who hath sworn that King James II is the only lawful king of this realm, or that he will bear faith and true allegiance to him, his heirs and successors, take those oaths to an usurper?' wrote Sancroft, in *The Present State of the English Government Considered, January, 1688–9*. The only occasion on which Sancroft addressed William of Orange was when James, having sent the Queen and infant Prince of Wales to France, left London in secret to follow them. Sancroft and the other bishops who were at the Guildhall meeting signed an address, deeming the assumption of government by William would be preferable to anarchy. Sancroft, however, showed his loyalty to James, by being one of the first to welcome him on his return to Whitehall.[62]

Despite his claim to a general delicacy of health, Sancroft regularly attended prayers in the chapel at Lambeth at six in the morning, twelve noon, three in the afternoon and nine at night. Nevertheless, in the midst of the upheavals of the Revolution, when Sancroft was elected Chancellor of the University of Cambridge, he declined, given his age and infirmities and recommended Lord Clarendon as more suitable. The University kept the post vacant for upwards of two months, hoping to prevail on him to accept it, but he was inflexible. Sancroft also refused to consecrate Gilbert Burnet, Bishop of Salisbury, which offended Burnet.[63]

To friends who warned him about the future Sancroft replied, 'I can live on fifty pounds a year', referring to his patrimony in Suffolk. William, in the light of the bishops' trial, was only too well aware of the affection of the people for Sancroft, and hesitated to proceed to deprive him of his diocese, and initially only suspended him from his office on 1 August 1689. But, during William's Irish campaign against James, Queen Mary, finding Sancroft at the end of six months determined not to take the oaths of allegiance, carried out the sentence of deprivation on 1 February 1690. At the same time, she deprived Turner Bishop of Ely; White of Peterborough; Lake of Chichester; Ken of Bath and Wells; Lloyd of Norwich; and Frampton of Gloucester, for the same offence. Bishops Lloyd of St Asaph, and Trelawny of Bristol were the only two of the seven prelates committed to the Tower who transferred their fealty to William and Mary. Dr Beveridge, who was nominated to replace Ken at Bath and Wells, asked Sancroft's advice how he should act. 'Though I should give my advice, I do not believe you will follow it', replied Sancroft. Beveridge assured him that he would. 'When they come to ask, say *nolo*; and say it from the heart. Nothing is easier than to resolve yourself what is to be done in the case.' Accordingly, Beveridge refused the see.

Sancroft did not acknowledge the authority of William and Mary, ignored the sentence of deprivation, and remained at Lambeth Palace, exercising his usual hospitality and charity. He said 'that he had committed no crime that could justly cause his degradation; so if the Queen wanted his house at Lambeth, she must either come, or send, and thrust him out of it by personal violence; for leave it in obedience to her mandate he would not'. His vacant see was filled by John Tillotson in April 1691. But Sancroft remained at Lambeth. Realising, however, that it would be necessary for him soon to leave, Sancroft told his chaplains, Needham and Wharton, the time had come for them to go. Sancroft also told them that 'a successor to my benefice is now appointed, and I can do you no more good, while it may be both prejudicial and dangerous to yourselves if you remain in my service'.[64] Wharton and Needham had both taken the oaths to William and Mary, and Sancroft did not wish to be a barrier to their preferment.

On 20 May, Sancroft received an order from the Queen to leave Lambeth Palace within ten days; he ignored it, and the process of ejection by law was begun. He was cited to appear before the barons of the Exchequer on 12 June, to answer a writ of intrusion. Sancroft's attorney avoided entering any plea which acknowledged William and Mary as sovereigns, or recognised their title. Judgement, of course, was passed against Sancroft and he left Lambeth Palace the same evening. He took a boat from Lambeth to the Temple, where he went to a house called the Palsgrave's Head, near Temple Bar, where he had previously sent his books and papers. He remained there for about six weeks without servants.

Sancroft left London on 3 August 1691, for Fressingfield, Suffolk. He refused the courtesy of having his correspondence transmitted free of charge through the secretary of state's office, or franked by any of the government officials. Despite the sympathy and accommodation of the new regime, Sancroft said, 'the spirit of calumny, the persecution of the tongue, dogs me even into this wilderness'.

In March 1691, Baptist Levinz, Bishop of Sodor and Man, although he had taken the oaths to William and Mary, wrote of his respect and admiration for Sancroft:

> I had, ever since I had the honour to know you, a very high veneration and respect for your grace, nor is my value at all lessened for you by the diminution of your fortunes. Calamity is but the touchstone of your virtues, and through this cloud your sincerity, your constancy, and other excellent endowments shine the brighter, and thereupon heighten my esteem for the most pious and admirable owner of them. I dare say no more, for your grace's modesty permits it not; yet still give me leave to love and honour you, and, as an abundant compensation, be pleased to bestow your benediction upon.[65]

The record of Sancroft's death, in November 1693, is preserved in the parish register of Fressingfield, which also contained the record of his birth. The assertion of Burnet, of his having raised a large estate out of the revenues of Canterbury, and left it to his family, was fully disproved by George D'Oyly.[66]

Schism and expulsions

The reluctance of the bishops to vote for the vacancy of the throne, and the accession of William and Mary as monarchs, caught some by surprise. The anonymous author of *Several Queries relation to the Present Proceedings In Parliament; more especially recommended to the Consideration of the Bishops*, complained of 'all the Disturbances which the Bishops shall make in the House of Lords', and asked 'whether the House of Lords will suffer themselves any longer to be imposed upon by the bishops in a thing that will be so injurious to the nation, as...not to comply with the House of Commons'. It advised the bishops to assist the nobility in the 'great work' of William's election.[67] One writer dredged up examples for the bishops to salve their consciences, but was forced to turn to the exclusions of Sigismund III of Sweden and Henry IV of France.[68] Another tract used the example of the removal of Mary, Queen of Scots from the throne of Scotland as further justification that the expulsion of James was lawful and valid for the bishops.[69]

Despite his comment to Lord Clarendon, Bishop Lloyd made his appearance on the platform at Westminster Abbey as one of the few bishops who assisted at the coronation. As joint monarchs, William and Mary were both crowned by Compton, but three quarters of the bishops resolved not to assist at the coronation. Lloyd performed the duty of the absent Archbishop Sancroft at the moment of recognition, by presenting Queen Mary to the people. Lloyd also preached a sermon on the anniversary of Gunpowder Plot on 5 November 1690, before the King and Queen, commemorating the landing of William at Torbay, which he treated as the climax of all the marvellous deliverances of the Church of England from popery. He received, as a reward, the office of Lord Almoner to King William.

In the remaining months before the non-juror bishops were turned out of their dioceses for refusing to swear oaths of allegiance to William and Mary, some loopholes were found for them. When, at the Lent sermons of 1689, some of the non-juring clergy were due to preach before the court, Gilbert Burnet permitted the prayers for the royal family to refer to the King and Queen, not clarifying whether it was William and

Mary, or James and Mary Beatrice, for whom they prayed. Bishop Turner called this a 'poor miserable expedient'.[70] Nevertheless, the bishops remained heroes in the literature of the Revolution. *The Confinement of the Seven Bishops*, published early in 1689, reminded readers:

> The charms of piety are no defence
> Against the new found power that can dispense
> With laws to murder innocence:
> Surely unless some pitying God look down
> And stop the threatening torrent, it will drown...
> The Bishops prisoners are, we tamely see
> The reverend prelates forc'd to bow the knee
> To Anti-Christ: no mighty monarch know
> Tho' we must pay to Caesar what we owe
> There is a power Supreme, by which you live
> Whose arm is longer and prerogative
> Larger by far, than yours, whose every word
> Can blast your hopes and turn your two-edged sword...
> Arise then, mighty sir, in God-like mean
> As of thy valour, let thy truth be seen
> Free from mistrust, let all your words be clear
> By action; let your promises appear
> Protect the Church, which brought you to the Crown
> You know 'tis great and honourable to own
> A kindness done; but to reward with death
> The happy instruments, that gave you breath
> Is mean; and mighty a Catholic conscience sting
> To cut the Hand of that anoints you King.[71]

Some clergy also debated the issue of resistance or passive obedience in the wake of the Revolution. In *A Friendly Debate between Dr Kingsman a dissatisfield clergyman and Gratianus Trimmer, a Neighbour Minister*, 'Dr Kingsman' asserted that if the Church had been dissatisfied with the King's policies, it should have adopted the usual way of dealing with such concerns, namely 'petitions, prayers, patience and tears'. 'Trimmer' replied

> As for petitions, you know the King sent the bishops to the Tower for an Answer, and thence brought them to the Bar. A warning to petitioners! Prayers were used, by such as you know, rather to harden than to soften the King's heart. Was he not commended to God, still

as his chosen servant? Was he not pray'd for, as if he had worshipped God in the best and only way? As for patience, it was exercised to the last day of safety. And as for tears, we durst not shed them for the King, nor for ourselves under him, for by Innuendo's they had been seditious.[72]

After the Convention, Bishop Lake retired to his diocese, refusing to transfer his homage to the new sovereigns. This refusal made for great popular animosity toward him. He knew that, if he persisted, his suspension from the diocese would take place on 1 August, and his deprivation would follow on 1 February 1690. He said, 'no matter, I will not take oaths which my conscience condemns. The hour of death and the day of judgment are as certain as the 1st of August and the 1st of February'. August came, and Lake was suspended; but three weeks later he had shivering fits, accompanied with convulsions and a fever. When his physicians, who saw the alarming nature of his symptoms, administered very strong and painful remedies, Lake smiled at their kindness. 'And is life worth all this at threescore years and five?' He died on 30 August 1690, his death was marked by an elegy, which included the lines:

Condemned to silence and to martyrdom
Because the world he'd learned to overcome

The lines were marked with an asterisk, which read 'when prisoner in the Tower'. Later in the elegy, Lake was called 'a candlestick of Gold' – one of the key images that was published in the wake of the trial.[73]

Lake's death was a relief to William and Mary, but they faced fierce opposition from Francis Turner. William and Mary had observed the absence of Bishop Turner from their coronation, and knew he did nothing to conceal his loyalty to James. When William departed for Ireland, Mary ensured that the oath of allegiance was tendered to Turner and the other non-juring bishops. It was unhesitatingly rejected by Turner, and his suspension from his diocese followed as a matter of course. 'On the last day of 1689 the Bishop of Ely was with me', wrote Lord Clarendon, 'and told me that a few days since, the Bishops of London and St. Asaph had been with my lord of Canterbury, pressing to know what he and the rest could do to prevent himself and the others being deprived. Could they make no steps towards the government?' Turner replied, he could do nothing 'if the King thinks fit, for his own sake that we should not be deprived, he must make it his business to devise expedients. We cannot vary from what we have done'.[74] Bishops

Compton and Lloyd were commissioned to tell Turner and the other non-juring bishops, that, if they would only remain quiescent, William and Mary would not proceed to enact the law against them, and would not appoint successors to replace them in their dioceses, but would leave them in quiet possession of their revenue and dignities. However, the bishops rejected this attempt at conciliation. The sentence of deprivation followed in February 1690. Turner protested against the validity of this sentence in the marketplace at Ely, and continued to preach in his robes every Sunday, in the chapel of Ely House, Hatton Garden, his London residence. Turner's sermons were attended by crowds, among whom was the Queen's uncle, the Earl of Clarendon, who always sat in a conspicuous place.

In the absence of William in Ireland, Queen Mary thought it prudent, instead of taking active measures against Turner, to send Bishop Lloyd to tell him privately, 'that their Majesties having been informed of the great resort of people to his chapel, were highly displeased, and he advised him therefore, as a friend, to shut it up for the time to come'. Turner did not submit, until a second indication of the risks he ran by flouting the law. 'On Tuesday, February 11', noted the Earl of Clarendon in his diary, 'the Bishop of Ely dined with me. He told me that "the Bishop of St. Asaph had been with him again, and told him plainly he must let no more company come to his chapel" so that I perceive all people are to have liberty of conscience, but those of the true Church of England.'

Turner regretted that James had fled the country. He was moved when he read the letter that the fleeing monarch left on the table at Rochester, explaining why he felt he had to leave England. From that moment, Francis Turner laboured to effect a counter-revolution.[75] Turner wrote confidentially to Sancroft, in January 1689,

> We came home from Lambeth four bishops in my coach, and we could not but deplore our case that we should disagree in anything, and such a thing as the world must needs observe. But their observing this, and insulting thereupon, makes it necessary for us in our own vindication to find out something on which we can agree.

He told Sancroft that there was to be a meeting that afternoon at Ely House of the leading clergymen, to discuss what was to be done.

> I enclose to your grace another paper which ought to be kept very private, but may be published one day, to show that we have not been wanting faithfully to serve a hard master in his extremity.[76]

Leading churchmen who had accepted William and Mary, met their non-juring brethren on a number of occasions at Lord Clarendon's house, and discussed public affairs. Thomas Tenison conceded that there had been irregularities in the settlement of the government, but the bishops should now make the best of it for fear of worse. Bishop Lloyd said 'it was known while things were in debate he had voted against abdication and for a regency, but now things being as they are, and the Prince of Orange crowned King, he looked upon acquisition to be just right'. Upon which, Clarendon replied, 'If you preach such doctrine it must not be to me'. Turner expressed regret that he and his six brothers had carried their resistance to James so far, and regretted they had not entered into recognisances for each other, instead of provoking James to send them to the Tower.[77]

Queen Mary showed her mettle when she arrested her uncle, Lord Clarendon, a key supporter of the non-juring bishops, and committed him to the Tower. Turner visited Clarendon, on 18 July, but was only permitted to see him in the presence of a warder. Turner came again on two further occasions, but was told that the Queen had expressly forbidden him access to Clarendon.[78] Subsequently, a plot against the government and to assassinate William was discovered, in which Turner was implicated by two letters, which were found among Lord Preston's papers, when he was arrested. The letters were addressed to James, under the pseudonym Mr Redding. The crown lawyers strove to prove that not only Turner, but all the non-juring bishops, were implicated in the plot to restore James and his family. The letter included what appeared to be incriminating lines from Turner: 'I speak in the plural, because I write my elder brother's sentiments as well as my own and the rest of the family; though lessened in number, yet if we are not mightily out in our accounts, we are growing in our interest that is in Jesus.' It was assumed by this that Turner meant that he spoke on behalf of Sancroft and other bishops. There was no proof that these incriminating letters were written by Turner; but they were a pretext to issue a proclamation for his arrest. He was fortunate to escape to the continent. Burnet recalled 'that the discovery of this correspondence gave the King a great advantage in filling the vacant sees'.[79]

There is no doubt that, after Francis Turner's return to England in 1691, he carried on a secret correspondence with James's exiled court at St Germains, and was involved in Sir John Fenwick's plot to assassinate William. When Fenwick was arrested, the government also pursued Parson Grascomb, a non-juring clergyman, who was the most active of all the pamphleteers who stirred up insurrection early in the reign of

William and Mary. At last he was discovered in the house of a French silk-weaver, in Spitalfields. The King's messengers surrounded the house with an armed force and captured Turner, among others. When questioned, the bishop said, very coolly, 'that he had no other account to give, but that he came there to dine, for he did not live there, his lodgings were at Lincoln's Inn'. When he found that the guards meant to detain him, he wrote to Secretary Vernon demanding his freedom, alleging 'that he held a pass to go to France if he chose, but he had made no attempt to avail himself of it'. Vernon referred Turner to Sir William Trumbull. Turner knew, as well as they did, that Shrewsbury, a leading minister, was also deeply involved in the plot, and was waiting for the right time to declare for James. Consequently, Trumbull left Turner at liberty. Turner retired to his lodgings in Lincoln's Inn, where he remained, apart from occasional visits to Moorpark, the home of Sir William Temple. In 1699, Turner moved himself and his daughter to a small house in the country. All his furniture from the episcopal palace at Ely had remained under the care of his brother, who was Principal of Corpus Christi College, Oxford.

On 14 October 1701, Turner wrote in code of 'the birthday of my unfortunate royal master, who now writes 68'. When he wrote this, Turner did not know that James had already died. Turner did not survive more than three weeks, dying on 2 November 1700. He was buried in the parish church of Therfield. In compliance with his desire, his remains were deposited by the side of his wife, his only memorial being the word 'Expergistur' – I shall awake.

If Turner was the avowed Jacobite agent among the seven bishops, Thomas White was responsible, most of all, for the non-juring church that continued after the Glorious Revolution. John Johnson, vicar of Margate, preaching the anniversary sermon for King's School, Canterbury in 1716, listed the names of the distinguished men who had attended the school, and observed,

> the memorable Thomas White, afterwards Bishop of Peterborough, was a scholar here, and I need not tell you that he was one of those seven prelates who made so notable a stand against arbitrary power in the year 1688; and yet afterwards, by his conduct, made it appear that his love to English liberty had not at all tainted the affection which he bore to his own natural lord and sovereign.[80]

After the Revolution, White sought to avoid all political excitement and occupied himself in his episcopal duties. It was assumed that White, as a chaplain of Princess Anne, would remain committed to his patron; but he refused to take the oaths of allegiance to the new sovereigns, so he was suspended and finally ejected from his bishopric and deprived of all his other preferments on 1 February 1690.

In the 1690s, Thomas White played a leading part in creating the non-juring church of England, hosting the consecration of a new non-juring bishop (nominated by the exiled James II) in February 1694. White assisted the deprived Bishops of Norwich and Ely, in consecrating Thomas Wagstaff, the non-juring Chancellor of Lichfield, and ejected rector of St Margaret Pattens, to the office of suffragan Bishop of Thetford. White performed the ceremony at Southgate, in the presence of Lord Clarendon, the Queen's uncle.

White was also rumoured to have composed the statement made by the Jacobite plotter Fenwick protesting his loyalty to James, along with his refusal to countenance violence against William. White expressed disgust that Tenison and Burnet led the bishops in voting for the attainder and execution of Fenwick, when by tradition bishops did not vote for the death sentences of those tried in the Lords. White later attended Fenwick on the scaffold. He died in London on 29 May 1698, his funeral and burial at St Paul's, on 4 June, provided an opportunity for non-juring Anglicans to commemorate a martyr both to divine right principles and to defence of the church against a King.

After the Convention, Bishop Thomas Ken retired to his diocese, to avoid taking the oaths to the new sovereigns. Although Ken had resolved not to abandon the oath he had sworn to James, his intentions were suspected. This was, in part, because some regarded him as time-serving. Even his oldest friend, Francis Turner, feared that the influence of George Hooper, the rector of Lambeth, who was a great friend of Ken, might persuade him to take the oaths to William and Mary. Turner expressed this in a letter to Sancroft:

> I must needs say the sooner we meet our brother of Bath and Wells the better; for I must no longer in duty conceal it from your grace (though I beseech you to keep it in terms of a secret) that this very good man is, I fear, warping from us and the true interest of the Church toward a compliance with the new government. I received an honest letter from him, and a friendly one, wherein he argues wrong, to my understanding, but promises and protests he will keep himself disengaged till he debates things over again with us, and that he was coming up for that purpose. My lord Bishop of Norwich has seen such another letter from him to my lord of Gloucester.

And upon the whole matter, our brother of Norwich, if your grace thinks fit, will meet us on Saturday; and I must needs wish my lord of Chichester would be there to help us, if need be, for it would be extremely unhappy should we at this pinch lose one of our number. I apprehend your parson of Lambeth has superfined upon our brother of Bath and Wells, and if he lodges again at his house I doubt the consequence: for which reason I will come over on Saturday morning to invite him to my country house.[81]

Evidence of Ken's equivocation appeared when a report was circulated that James had enacted a deed which transferred Ireland to the King of France. Ken, in surprise and indignation, declared his intention, if this was true, to take the oaths to William and Mary, and suggested the clergy in his diocese do the same. But, on Ken's arrival in London, he found the rumour was false, and he burned the paper proposing to take the oaths to William and Mary. Instead, he persevered in risking deprivation rather than abandon his allegiance to James.[82] Gilbert Burnet, who had recently been appointed to the bishopric of Salisbury, wrote a letter to Ken on his refusal to take the oaths:

I am the more surprised to find your lordship so positive, because some have told myself that you had advised them to take that which you refuse yourself, and others have told me that they read a pastoral letter which you had prepared for your diocese, and were resolved to send it when you went to London. Your lordship, it seems, changed your mind there, which gave great advantage to those who were so severe as to say that there was something else than conscience at the bottom. I take the liberty to write thus freely to your lordship, for I don't deny that I am in some pain till I know whether it is true or not.[83]

Ken calmly replied, on 5 October 1689, that people in his diocese knew his views on the oath. He also said that, 'when, on his arrival in town, having discovered the incorrectness of the statement, he had burned the paper, and adhered to his original determination'. This, he said, was not a change of mind, and that he would not take lessons in changing minds from people who had switched their allegiance to William. He added that a good conscience was better than the trappings of office, and he cared nothing for the ridiculing of his opponents. 'His most holy will be done; though what particular passion of corrupt nature it is which lies at the bottom, and which we gratify in losing all we have, it will be hard to determine. God grant such reproaches as these may not revert on the authors.'[84]

The letter was written at the house of his old friend, Hooper, who had taken the oaths to William and Mary, and tried to persuade Ken do so too. Ken silenced him with these words:

> I question not but you and several others have taken the oaths with as good conscience as I shall refuse them, and sometimes you have almost persuaded me to comply by the arguments you have used; but I beg you to use them no further, for should I be persuaded to comply, and after see reason to repent, you would make me the most miserable man in the world.

As Ken persisted in his refusal to take the oaths, he was served with a writ of ejection. The see of Bath and Wells was first offered to William Beveridge, who followed Sancroft's advice, and refused it, and then accepted by Richard Kidder. Ken initially retired to the house of his nephew, Isaac Walton, prebendary of Salisbury, until a permanent home was offered by his former college friend, Thomas Thynne, Viscount Weymouth, in a suite of apartments in the upper storey of Longleat, Wiltshire.

The government regarded Ken with some concern, and watched for occasions to act against him. The only pretext they could find for prosecuting him was for his charitable collections for the ejected non-juring clergy, for which Ken was cited to answer to the Privy Council. He immediately surrendered himself, and, despite ill-health, presented himself, in his patched and threadbare episcopal dress, at the Privy Council, on 28 April 1696.[85] He was asked whether he had subscribed to the charity fund for the needy. 'I thank God I did', replied Ken;

> and it had a very happy effect, for the will of my blessed Redeemer was fulfilled by it, and what we were not able to do was done by others; the hungry were fed, and the naked were clothed; and to feed the hungry, to clothe the naked, and to visit those who are sick or in prison...[86]

The deprived Bishop of Norwich, William Lloyd,[87] wrote a letter to Ken, on the death of William III, asking him to come to London, to consult with their non-juring brethren. But Ken, who was ill, and wanted to avoid all political agitation, replied to Lloyd:

> that his counsel and assistance were not worth a London journey, which was consistent neither with his purse, his convenience,

health, nor inclination; that he had quite given over all thoughts of re-entering the world, and nothing should tempt him to any oath; but he heartily desired an expedient could be found for putting an end to the present schism.[88]

In 1703, Queen Anne sent for George Hooper, and told him she intended the see of Bath and Wells for him, but he asked her to restore it to Bishop Ken. She thanked Hooper for mentioning it, and told Ken he might return to his see without swearing any oaths. Two great obstacles to Ken's resumption of his diocese had been removed: James II and William III were both dead. Yet he firmly declined her offer, claiming that he was too old and ill. Hearing, soon after, that it had been offered to his friend, Hooper, then Bishop of St Asaph, and that Hooper had refused it, he wrote:

> I am informed that you have had an offer of Bath and Wells, and that you have refused it, which I take very kindly, because I know you did it on my account; but since I am well assured that the diocese cannot be happy to that degree in any other hands than in your own, I desire you to accept of it, and I know that you have a prevailing interest to procure it. My nephew and our little family here present your lordship their humble respects, and will be overjoyed at your neighbourhood.
>
> I told your lordship long ago at Bath how willing I was to surrender my canonical claim to a worthy person, but to none more willingly than to yourself. My distemper disables me from the pastoral duty, and had I been restored, I declared always that I would shake off the burden and retire. I am about to leave this place,[89] but if need be, your archdeacon can tell you how to direct to me.[90]

After the translation of Hooper to Bath and Wells, Ken discontinued his episcopal style, and now signed himself 'Thomas, late Bishop of Bath and Wells'. Queen Anne, on the suggestion of the lord treasurer, Godolphin, settled a pension of £200 per annum on Ken. He died on 19 March 1711, aged 74, and bequeathed all his books to his friend and benefactor, Lord Weymouth.

The two survivors of the seven bishops were William Lloyd and Jonathan Trelawny, both of whom embraced the Revolution in a way that Sancroft, White, Lake, Turner and Ken would not. Surprisingly, Lloyd was absent from the Convention vote on conferring of the crown on William and Mary, perhaps, as Michael Mullett claimed, out of equivocation.[91] However, he attended the coronation and was appointed almoner to the Queen. He laboured hard to keep the non-jurors within the Church, but, in the end, agreed to consecrate Tillotson as Sancroft's successor as archbishop of Canterbury. Lloyd was also responsible for the stoutest Anglican defence of the Revolution, *A Discourse of God's Ways of Disposing of Kingdoms*, published in 1691. In 1692, he was rewarded with preferment to Lichfield and Coventry and, in 1699, to the diocese of Worcester. In the latter diocese, his reputation, as a Whig bishop brought him into conflict with both Dissenters and Tory Jacobites, and his interference in the election of 1702 was investigated by the House of Commons. In his last decade, he devoted himself to administering his diocese and to arcane theological research.

Jonathan Trelawny also gathered the reward of his part in revolutionary affairs. He was appointed to the diocese of Exeter three days after the coronation of William and Mary. Trelawny became something of an oddity; a Tory churchman who had supported the Whig Revolution of 1688. At Exeter, he used the church courts to impose episcopal discipline and, in 1704, he even advocated that Convocation should censure bishops for neglecting their disciplinary powers. He was a fierce opponent of developments such as the societies for the reformation of manners, which he felt undermined the Church of England's role as the national Church, and therefore the guardian of its morals. In 1706, he was appointed to Winchester, through the influence of the Whig Junto ministry of Marlborough and Godolphin, his friends from the Revolution. Unlike most Tories, he had strongly supported the war with France, and did not easily accept the Peace of Utrecht. Some saw him as a time-server and a vicar of Bray, whose principal value for the government was as a borough monger. In fact, Trelawny was a dedicated, if stern, bishop and was at pains to defend the Church whenever he felt it was threatened. He died in August 1721, 33 years after his acquittal by the judges in the summer of 1688.

Conclusion

What light does the trial of the seven bishops cast on the events of 1688–9? It is clear that the bishops were fearful of James's policies and distrusted his objectives and tactics. They clearly felt that James had broken his coronation oath, that he was threatening the Church, and had arrogated to himself an illegal and general dispensing power to do so. Doubtless they felt grievously disappointed by James, whose succession they had championed. The Magdalen College affair had demonstrated the ineffectiveness of traditional Anglican passive obedience; Hough and the fellows had failed to halt James's illegal actions. Indeed the fellows were only reinstated after the bishops had challenged James and been acquitted. Lay politicians stood by in equivocation. Allowed to continue, it seemed likely that the King would seriously damage both Church and state. Yet the bishops' attitude remained essentially conservative and can be seen in the words of a speech that Bishop Lloyd intended to deliver (but did not have a chance to) during the trial. He wrote of their behaviour:

> We, conspire against the King? – to undermine his Government by making and publishing seditious libels? What strange kind of men they would make of us! We, that not only by past obligations, but by our present interests, and all our hopes of this life and the future, have not only the strongest, but all possible motives to hold us to obedience and loyalty? In a word, we, whose holy Religion teaches us, under pain of damnation, not to rebel against our King, though he be of another religion; nay thought he should be an enemy to our religion?... We do not see that we have transgress'd the bounds of our duty.

They abhorred James's first Declaration, 'yet all that twelve month we were silent; we laid out hands on our mouths...'[1] Lloyd was perhaps over-egging the justification but, in essence, what he planned to say was a genuine reflection of the bishops' position; they did not believe they had conspired to be seditious, they had not rebelled against the civil power, they had not resisted lawful authority – indeed they had been passively obedient for too long. But they were willing to reject unlawful government and resist a tyrant who was threatening the Church.

There the consensus ended. As far as the way forward, Lloyd and Trelawny – after the acquittal – had crossed the Rubicon and were committed to ejecting James; this was because they could not conceive of James as a ruler they could trust. Sancroft, Turner, Ken, White and Lake were content to support him, if James would stand by his commitment to the 'ten advices' they had forced on him in the autumn of 1688. Thus a conservative Anglican revolution, perhaps with an element of the toleration of Dissent, was acceptable to the majority of the bishops. But, by opposing James, and getting away with it, they had set in train events which could not be controlled. The reality of the situation was, as Straka claimed, 'from the Summer of 1688 and through the Winter of 1689, the Church found itself less and less able to cope with a revolution that it had to a large degree started'.[2] Others had seen in the bishops' petition, imprisonment, trial and acquittal that not only was James in error, but that he did not enjoy popular support and could be resisted. The aristocracy, the politicians and those who had disengaged from James and stood on the sidelines, the man and woman in the pew, the parson in his pulpit – all of them knew it, and that knowledge gave them the courage and strength they had lacked before.

It is difficult to look at the trial of the bishops without indicting James. The revisionists who have defended him are perhaps right that such an approach is unhelpful. But, equally, the revisionists have given James the benefit of the doubt that fails to take into account either James's own character, or the circumstances in which his opponents found themselves. The problem for James's revisionists is that even those to whom he was closest did not trust him. In January 1689, Sunderland, admittedly not one of James's most truthful or consistent loyalists, published *The Earl of Sunderland's Letter to a Friend*, which argued that he had tried to advise James well but that James was headstrong and refused good counsel. Certainly this is consistent with Sunderland's dealings with James on the packing of Parliament, the postponement of elections and the legal action James took against the seven bishops.[3] Nevertheless, Sunderland's role, as J. P. Kenyon asserted, is open to the question; if James was so impervious to his advice why did Sunderland remain? But this denied the Cavalier Tory optimism, which Sancroft also shared. Even to the end, the King's party – including some of the seven bishops – hoped against hope that James would change, that he would become the ruler they wanted him to be. In this, hope was truly disappointment deferred. But, in the final analysis, the Tories left James because of his treatment of the Church.[4]

A further problem for revisionists is that James began his reign by seeking to crush the Dissenters entirely.[5] James's track record was damning. His persecution of Richard Baxter in May 1685 shocked the Dissenters. In Scotland, James's action against Margaret Maclachlan and Margaret Wilson, both of whom were sentenced to drowning for their refusal to abandon the covenant and accept episcopalianism, a sentence James refused to commute, was similarly brutal. James's action against those who preached against Catholicism indicated his contempt for toleration. Besides the action against John Sharp, James refused William Sherlock his pension as master of the Temple on the grounds that he had preached against Catholicism. Contemporaries also noted the record of Catholic rulers in treating their Protestant subjects with brutality. During the Revolution, the behaviour of the Catholic rulers of Spain, Portugal and France was atomised, including 'the persecutions of the Protestants in the vallies of the Piedmont'.[6] *The Hard Case of Protestant Subjects under the Dominion of a Popish Prince*, published during the crisis, referred to the death penalty which Catholic canon law imposed on those judged heretic, and one which no secular judge could overrule.[7] In *Ten Seasonable Queries*, written by an Englishman in Amsterdam, it was asked 'whether any real and zealous Papist was ever for Liberty of Conscience, it being a fundamental principle of their religion that all Christians that do not believe as they do are hereticks and ought to be destroyed'.[8]

It is also clear that James, as one contemporary put it, was complacent in relying on

> the Non-Resisting doctrine [that] had so ty'd the hands of the Church of England men, that they [James and his friends]...might ridicule the Bond that bound us to our good behaviour. The Dissenters were, as they thought, so obliged by the Liberty of Conscience, and the fulsome Applications they made to them in many ill writ pieces.[9]

The dilemma for the Cavalier Anglicans was that they had supported James and his cause against the exclusionists, who were friendly to the Dissenters. But, as Ryle argued, there was a community of interest between the Anglicans and Dissenters, 'they [the bishops] supplied unanswerable proof that the real, loyal, honest, old-fashioned High Churchmen disliked Popery as much as any school in the Church'. Ryle credited the bishops with the primacy of Protestant principles:

> There is an overwhelming mass of evidence to prove that the real reason why the Seven Bishops resolved to oppose the King, was their determination to maintain the principles of the Reformation and to oppose any further movement towards Rome.[10]

The Protestant ideology, which formed the basis for the bishops' petition, was rooted in a belief that, from the Reformation, a persecuting form of Catholicism was determined to reclaim England. The Protestant martyrs of Mary's reign, and the triumph over the Spanish Armada were vivid, if distant, memories, kept alive by repeated reprints of Foxe's *Book of Martyrs*, and much of the anti-Catholic propaganda, during the period of the trial, was focused on the persecuting nature of Catholicism.

The author of *The Reasons of the Suddenness of the Change in England* grappled with the reasons why James was so popular in 1685 but ejected in 1688. The core of it was James's insincerity and willingness to say anything to get his own way.[11] When James met leading Dissenters, in October 1688, and promised them that he was still committed to liberty of conscience, they knew either he was lying to them, or he had been lying to the bishops, with whom he had recently been reconciled. Both, of course, knew James was completely untrustworthy.[12] The nub of the matter was that James and his supporters were

> equally mistaken in their carriage towards the Church of England party, for when some of them had pursued both Clergy and Laity with the utmost obloquy, hatred, oppression and contempt to the very moment they found the Dutch storm would fall upon them. Then all at once they passed to the other extream, the Bishops are presently sent for, the government intirely to be put into their hands, and all places, presses and papers fill'd with the Encomiums of the Church of England's Loyalty and Fidelity, who but three days before were malecontents, if not rebels and traitors for opposing the King's Dispensing Power...[13]

From this position there was no return. If neither James's supporters, the aristocracy, the Dissenters nor the bishops – or at least most of them – believed James would keep his word and honour his undertakings, there was little that could be done. This seemed to be implicit in the suggestion of a regency during the Convention, which few but the future non-juror bishops were willing to consider.

Whereas Sancroft came late to the scramble to win Dissenters to the Anglican cause, he had been well-served by earlier Anglican moves to show Dissenters that the Church of England was not their enemy.[14] This ingredient in the events of 1688 should not be underestimated. If James had been able to win over popular support from Dissenters, if they had ignored Halifax's *Letter to a Dissenter* and the mass of Anglican propaganda, and welcomed James's Declaration, in the same way that they had greeted Charles II's Declaration of Indulgence in 1672, events might have been very different. If, with Dissenting votes, James had secured electoral dominance over the borough seats and obtained a majority in the Commons for repeal of the Test Act, all might have been lost for the bishops. All might also have been lost for William too, if the army and navy had been placed under Catholic officers, which would have been a consequence of repeal.

To move to broader questions, was the Revolution which the trial of the seven bishops sparked a coup d'etat or a popular revolution? It is clear that it was both. That Lloyd and Trelawny and others were in contact with William, and aware of the conspiracy to evict James, suggests that there was a willingness by the governing classes to be rid of James in a 'palace coup'. The actions of Danby, Churchill and Compton clearly indicate that there was a coordinated conspiracy against James. But the popular feeling in support of the bishops was an example of the widespread opposition to James, without which a coup, and even an invasion, might not have succeeded. When, as William moved across the West Country, the people, along with the gentry and aristocracy chose not to defend James they were making a clear choice, as much as when they had refused to join Monmouth's invasion. In 1685, the decision of the nobility and people not to support an invasion indicated a choice for James and legitimacy; in 1688, to support an invasion was a rejection of James and tyranny. In the wake of the bishops' imprisonment and trial, many people, great and poor, wrestled with their consciences and chose not to act. In other words, not acting, not defending James, was as much a positive choice as fighting for him. John Churchill, Charles Trelawny, Clarendon, Cornbury and the rest made a choice to abandon James as clearly as the populace that refused to fight for him. This was not a revolution that took place in the absence of expressions of popular opinion; popular opinion kept people at home, and turned its back on James. Thus the apparent choice of seeing the Glorious Revolution as a coup or a popular rebellion is a false choice, it was both and the two were interdependent; without the popular opposition, elicited by the trial, the conspirators would have been less likely to act.

The popular engagement with the Revolution was mediated through the publication of pamphlets and tracts. One of the clearest aspects of the events surrounding the trial of the bishops, and one that endorses the involvement of the wider population in the events of 1688, was the way in which both James and the bishops conducted propaganda campaigns of varying success to appeal to popular opinion. There was a material culture, and willingness to use it, that brought the Glorious Revolution into peoples' homes. The fact that both King and episcopate sought to ensure that they were portrayed in a particular light in the public press emphasises how important public opinion was. In almost all the cases, the bishops outpaced the King in their management of public opinion. Their canvassing of the opinions of the London clergy, and of London Dissenters in the week before 18 May 1688, was almost a model of modern public relations. In this, even Sancroft can be credited with a progressive approach. The meetings with influential clergy and Dissenters were used to dramatic effect in consolidating resistance to James's order that the Declaration should be read in churches. Yet this would not have been possible without a major campaign, begun in London by Bishop Compton, to ensure that Catholicism was demonised and that the Church was depicted as the one legitimate body to which all Protestants, Anglican or Dissenter, owed their loyalty.

There was also an important theological and apocalyptic aspect to the propaganda of James's reign. As Warren Johnston asserts

> a profusion of tracts, treatises, and sermons explaining the apocalyptic implications of the events of 1688–9...reached a wide audience and still informed political and religious opinion in late seventeenth century England. Conspicuous among these writings was a shared conviction among Anglican and Nonconformist authors that the Revolution marked a significant step towards the fulfillment of the overthrow of the papal beast, the fall of the Roman Babylon, and the destruction of the antichrist, as well as stressing England's central role in these apocalyptic achievements.[15]

After the Revolution, an engraving showed Louis XIV and James II pulling a rope attached to a windmill (made from the mitres of the seven bishops) and the pope, restrained by William III.[16] To add to the apocalyptic flavour of the times, Simon Patrick, in January 1689, claimed that 'the time is coming apace, when Christianity will end, as it began, in abundance of Truth and Peace'.[17]

Popular perception regarded the bishop's petition, imprisonment and trial as the pivotal moments of James's reign. The view of the bishops as heroes of the Revolution extended to the central role they played in displacing James. In a popular ballad of 1689, *A King or No King*, James was shown lamenting:

> I wish I had not sent to the Tower
> Those Seven Bishops me Counsel gave
> But now at last I am out of Power
> And you no more K. James shall have.[18]

Even those popular prints that blamed James's counsellors, such as Lord Chancellor Jeffreys, saw the bishops as central to the events. In *The Lord Chancellors Villanies*, the verse ran

> Then next to the Tower our Bishops was packt
> And swore he had done a very good act
> But now shall be tried for the matter of fact
> Sing hey brave Chancellor, O fine Chancellor,
> Delicate Chancellor oh!
> And when that the Bishops were brought to be try'd
> To accept a Petition they humbly desir'd
> He swore he would prove it a Libel he cry'd
> Sing hey brave Chancellor, O fine Chancellor,
> Delicate Chancellor oh![19]

Corroboration for this sense of the way in which the bishops were pathfinders for popular opinion came from those who had played an ambivalent role in the events of 1688. In the jostling among James's former supporters, that of Thomas Sprat, Bishop of Rochester, was the most instructive. Sprat was keen to defend his actions: he had served on the Ecclesiastical Commission and this indelibly associated him with James. In February 1689, he published a tract justifying his actions. Claiming he had been appointed to the Commission without his knowledge, and that he had not known that Sancroft had refused to sit on it, he said he was 'surprised' by the action against Compton, had voted for his acquittal and obstructed further action against him and against the universities. In a desperate claim, Sprat tried to appropriate some of the credit attached to the bishops:

> I might add moreover that it is very probable I was at last in more imminent danger than any of my brethren...the petitioning Seven Bishops, and my Lord of London only excepted; whose merits and sufferings in asserting our laws and religion were so conspicuous, and by consequence the fury of the Papists against them so implacable, that perhaps it would be presumption in any other clergyman, much more in me, to come in competition with them for either of those honours.[20]

In his *Second Letter*, Sprat claimed that 'what passed between King James and some of the Bishops a little before the late wonderful Revolution...had a considerable effect for the benefit both of Church and State in that critical time'. Sprat recognised that 'the Bishops had then as difficult a post to maintain, and maintain'd it as firmly as any other order of men in the Kingdom'.[21] Sprat argued

> 'Tis evident the Seven Bishops...had such an opportunity put into their hands by God's Providence, for the overthrow of Popery and Arbitrary Power, by their sufferings for delivering their sense of King James the Second's Declaration.[22]

Sprat claimed, somewhat implausibly, that, had he been given the opportunity, he would have supported their petition. He also claimed

> From that petition of those bishops, so defended by the invincible arguments of their learned council on that day; and so justified by the honest verdict of their undaunted jury on the next day; from thence, say I, we may date the first great successful step that way made towards the rescuing of our Laws and Religion.[23]

The role of the bishops was enshrined in the Revolutionary settlement. The first three crimes, listed by the Commons as justification for excepting persons from the proposed Bill of Indemnity in 1689, were asserting and promoting the dispensing power of the King, involvement in the commitment and prosecution of the seven bishops and advising and promoting the Ecclesiastical Commission. In the Bill of Rights the 'pretended power of suspending of laws' was also declared illegal.[24]

While the bishops had an honoured place in the new regime's view of the Revolution, the fact that the majority became non-jurors undoubtedly damaged this credit. The decision to abandon their offices and responsibilities in the new regime, rather than swear the oaths to William and Mary seemed to imply that they had second thoughts about ejecting James from the throne, and this ran counter to the popular mood. By the end of 1689, most of the seven bishops had thereby affected their standing in the popular mind. Some bishops seemed keen to see James return, and Turner was clearly opposed to anything other than a restoration of James. It was known that Bishop Lake had openly prayed for James at St Mary le Bow.[25] Some bishops opposed the concession of comprehension in 1689, which aborted much of the hopes for more toleration of dissent.

Were the bishops, and their role in the Revolution, an expression of a conservative desire for continuity, or a desire for change and progress? Here the answer is no less complex. The bishops clearly would have preferred the status quo, a Tory monarchy allied to the Church which excluded Dissent and Catholicism from any share in government or public life. Arguably, this is what they achieved in the autumn of 1688 with James's acceptance of their 'ten advices'. But, in the absence of the status quo, they sought a complicated resolution. At their trial, the bishops pointed to events in 1662, 1672 and 1685 when the dispensing power had been rejected; but they were not seeking a return to a period in which Charles II had also sought to tolerate religious difference, nor to one in which Charles had tried to use the dispensing power. Rather, they sought a return to the brief period, from 1679 to 1685, when the state had been dependent on the Church for the disciplining of Dissent and the King hoped to secure Parliament only for the Anglicans.[26] But to do so meant curtailing the ambitions and power of the King. Thus even Sancroft, Turner and the future non-jurors sought a reformed monarchy, and were probably prepared to make peace with the Dissenters as the price for it. Patrick Dillon wrote, 'the truth is that no ideology survived the Revolution intact'. Tories were monarchists who evicted their monarch; Whigs were supporters of toleration who did not achieve full religious tolerance.[27]

John Ryle wrote of the trial of the seven bishops:

> The importance of that event is so great, and the consequences which resulted from it were so immense... [but] attempts are made to misrepresent this trial, to place the motives of the bishops in a wrong light, and to obscure the real issues which were at stake... They [the bishops]

are of unspeakable importance. They stand out...in the landscape of English history, like Tabor in Palestine, and no Englishmen ought ever to forget them. To the trial of the Seven Bishops we owe our second deliverance from Popery.[28]

The misrepresentation of the bishops and questioning of their motives was a natural consequence of their breach with the Church in 1689, which they had claimed to defend in 1688. Indeed, to historians, the Church was sometimes seen as a victim of the Revolution of 1688.[29] By forming the core of the non-jurors, Sancroft and his friends obscured the central role they played in the events of 1688.

It is too strong to claim that there would not have been a Revolution without the bishops' petition, imprisonment and trial, but their role, by mobilising conspirators and public opinion, made the process a much easier one, and one which probably spared much bloodshed. It is clear that William would not have allowed a Catholic heir to replace him and Mary, and he had already decided on intervention in England. But without the events of May and June 1688, his task would have been much harder, less assured, and could have led to a second civil war. The seven bishops may not have been the progenitors of the Glorious Revolution, but they were its midwives.

Notes

Preface

1. M. Barone, *Our First Revolution, the Remarkable British Upheaval That Inspired America's Founding Fathers*, New York, 2007; G. S. de Krey, *Restoration and Revolution in Britain: A Political History of the Era of Charles II and the Glorious Revolution*, London, 2007; P. Dillon, *The Last Revolution: 1688 and the Creation of the Modern World*, London, 2006; T. Harris, *Revolution: The Great Crisis of the British Monarchy, 1685–1720*, London, 2006; S. Pincus, *England's Glorious Revolution*, Boston MA, 2006; E. Vallance, *The Glorious Revolution: 1688 – Britain's Fight for Liberty*, London, 2007.
2. G. M. Trevelyan, *The English Revolution, 1688–1689*, Oxford, 1950, p. 90.
3. W. A. Speck, *Reluctant Revolutionaries, Englishmen and the Revolution of 1688*, Oxford, 1988, p. 72. Speck himself wrote the petition 'set off a sequence of events which were to precipitate the Revolution'. – Speck, p. 199.
4. Trevelyan, *The English Revolution, 1688–1689*, p. 87.
5. J. R. Jones, *Monarchy and Revolution*, London, 1972, p. 233.

Introduction: The Seven Bishops and the Glorious Revolution

1. A. Rumble, D. Dimmer et al. (compilers), edited by C. S. Knighton, *Calendar of State Papers Domestic Series, of the Reign of Anne Preserved in the Public Record Office*, vol. iv, 1705–6, Woodbridge: Boydell Press/The National Archives, 2006, p. 1455.
2. *Great and Good News to the Church of England*, London, 1705. The lectionary reading on the day of their imprisonment was from Two Corinthians and on their release was Acts chapter 12 vv 1–12.
3. *The History of King James's Ecclesiastical Commission: Containing all the Proceedings against The Lord Bishop of London; Dr Sharp, Now Archbishop of York; Magdalen-College in Oxford; The University of Cambridge; The Charter-House at London and The Seven Bishops*, London, 1711, pp. 1, 9, 27.
4. *What has been may be: Or A View of a Popish and an Arbitrary Government... To which is added the Twenty-One Conclusions Further Demonstrating the Schism of the Church of England*, London, 1688.
5. *Bishop Sancroft's Ghost with his Prophesie of the Times, And the Approach of the Anti-Christ*, London, 1712, pp. 1–2.
6. M. G. Smith, *Fighting Joshua, Sir Jonathan Trelawny 1650–1721*, Redruth, 1985, p. 174.
7. Poem XCIII 'On the Seven Bishops' in *A Collection of Epigrams*, London, 1727.
8. National Library of Wales [NLW], Plas Yn Cefn Ms, 2939.

9. The second of these (*The Proceedings and Tryal in the Case of the Most Reverend Father in God, William, Lord Archbishop of Canterbury, ... in the Court of the King's Bench*, London, 1739) was printed 'for the booksellers in town and country'.
10. *The Eclectic Review*, April 1848, p. 557.
11. Ibid., pp. 549–51.
12. *Lords Hansard*, 30 June 2005, col. 356.
13. T. B. Macaulay, *History of England*, London, 1848, vol. 2.
14. The full title of Strickland's book was *The Lives of the Seven Bishops Committed to the Tower in 1688 Enriched and Illustrated with Personal Letters, Now First Published, from the Bodleian Library*, London, 1866.
15. H. Lucock, *The Bishops in the Tower*, London, second edition, 1896, p. xi.
16. C. Hill, *The Century of Revolution 1603–1714*, London, 1961, pp. 220–39.
17. M. Mullett, *James II and English Politics 1678–1688*, London, 1994, pp. 29, 49, 57.
18. G. M. Straka, *Anglican Reaction to the Revolution of 1688*, Wisconsin 1962, pp. 21, 120pp.
19. G. Every, *The High Church Tradition*, London, 1956, p. 21.
20. Published in 1985.
21. B. W. Hill, *The Growth of Parliamentary Parties 1689–1742*, London, 1976.
22. K. Wilson, 'Inventing Revolution: 1688 and Eighteenth Century Popular Politics' in *Journal of British Studies*, vol. 28, 1989, p. 350. However Wilson concedes that the Revolution came to be *understood* as revolutionary because it gave rise to radical ideas of popular and non-violent political change.
23. J. C. D. Clark, *English Society 1660–1832*, Cambridge, 2000, pp. 74, 80, 85. All of this falls into a section headed 'Church before State: The Revolution of 1688'.
24. J. R. Jones, *The Revolution of 1688 in England*, London, 1972.
25. Ibid., p. 12.
26. J. Stoye, 'Europe and the Revolution of 1688' in Beddard (ed.), *The Revolutions of 1688*, Oxford, 1991, p. 208.
27. See for example D. H. Horsford, 'Bishop Compton and the Revolution on 1688' in *Journal of Ecclesiastical History*, vol. xxiii, 1972, p. 209.
28. J. Miller, *James II a Study in Kingship*, Hove, 1977; J. R. Western, *Monarchy and Revolution: The English State in the 1680s*, London, 1972; B. Coward, *The Stuart Age: A History of England 1603–1714*, London, 1980.
29. L. Pinkham, *William III and the Respectable Revolution*, Cambridge MA, 1954.
30. J. Israel (ed.), *Anglo-Dutch Moment, Essays on the Glorious Revolution and Its World Impact*, Cambridge, 2003; J. Black, *A System of Ambition? British Foreign Policy 1660–1793*, London, 1991.
31. T. Claydon, *Europe and the Making of England 1660–1760*, Cambridge, 2007, p. 243.
32. Jones, *The Revolution of 1688*, pp. 7–17 and 328–31.
33. G. V. Bennett, 'The Seven Bishops: A Reconsideration' in D. Baker (ed.), *Studies in Church History*, vol. 15, 1978.
34. W. M. Spellman, *The Latitudinarians and the Church of England 1660–1700*, Athens GA, 1993, pp. 54, 134.

35. Roger D. Jones, 'The Invitation to William of Orange in 1688' University of Wales Cardiff, MA thesis, 1969, p. 31.
36. J. P. Kenyon, 'Introduction' in E. Cruickshanks (ed.), *By Force Or by Default? The Revolution of 1688–1689*, Edinburgh, 1989, p. 3.
37. J. Miller, 'James II and Toleration' in E. Cruickshanks (ed.), *By Force Or by Default*, p. 8.
38. Ibid., p. 19. As Miller conceded 'the suddenness of James's conversion from persecution to toleration is, I think, a significant argument against claims that he consistently favoured toleration'. But equally he asserted 'I have suggested that those who saw James's advocacy of toleration as an example of Popish duplicity were wide of the mark'.
39. Miller, *James II*, pp. 124, 126, 127. Did James have a plan to forcibly convert the English to Catholicism? Certainly Barillon, the French Ambassador, told Louis XIV that 'this prince has thoroughly explained to me his intentions with regard to the Catholics, which are to grant them entire liberty of conscience and the free exercise of their religion'. When, in April 1687, he issued his first Declaration of Indulgence he said 'we cannot but heartily wish, as it will be easily believed, that all the people of our dominion were members of the Catholic Church'.
40. Ibid., pp. 128, 143–5.
41. Ibid., pp. 64, 128, 155, 168, 196–7, 241.
42. J. Miller, *Popery and Politics in England 1660–1688*, Cambridge, 1973, pp. 258, 259.
43. Smith, *Fighting Joshua*, p. 44.
44. Sunderland was rehabilitated to some degree by J. P. Kenyon who argued that his position may not have been quite as cynical as previously assumed. Kenyon judged Sunderland as a maladroit and credulous royalist rather than any great schemer capable of betraying James. J. P. Kenyon, 'The Earl of Sunderland and the Revolution of 1688' in *Cambridge Historical Journal*, vol. 10, 1955.
45. A. F. Havighurst, 'James II and the Twelve Men in Scarlet' in *Law Quarterly Review*, vol. 69, 1953, p. 545.
46. R. A. Beddard, 'The Unexpected Whig Revolution of 1688' in Beddard (ed.), *The Revolutions of 1688*, Oxford, 1991, p. 97.
47. D. Defoe, *A New Test of the Church of England's Loyalty: Or Whiggish Loyalty and Church Loyalty Compar'd*, London, 1702, p. 16.
48. J. P. Kenyon, *The Nobility in the Revolution of 1688*, University of Hull, 1963, p. 7.
49. M. McCain, 'The Duke of Beaufort's Defence of Bristol during the Monmouth's Rebellion 1685' in *Yale University Gazette*, April 2003.
50. Kenyon, *The Nobility in the Revolution of 1688*, pp. 8–19.
51. J. Carswell, *The Descent on England*, London, 1969, p. 144.
52. At the same time, other historians have emphasised the Protestant motives of those active against James – even in Catholic strongholds. – P. D. Heitzman, 'The Revolution of 1688 in Lancashire and Cheshire', Manchester University MPhil thesis, 1984.
53. M. Goldie, 'The Political Thought of the Anglican Revolution' in Beddard (ed.), *The Revolutions of 1688*, Oxford, 1991, p. 105.
54. Ibid., pp. 113–21, 132, 135.

55. Ibid., pp. 122, 123, 133.
56. Ibid., p. 124.
57. Y. Sherwood, 'The God of Abraham and Exceptional States, Or the Early Modern Rise of the Whig/Liberal Bible' in *The Journal of the American Academy of Religion*, vol. 76, no. 2, 2008.
58. W. A. Speck, *Profiles in Power: James II*, London, 2002 and Harris, *Revolution: The Great Crisis of the British Monarchy*.
59. Harris, *Revolution: The Great Crisis of the British Monarchy*, pp. 14, 236, 240, 275–7, 478, 486–514. This view seems to endorse Dickinson's judgement that the scholarly consensus is that the Revolution was conservative. H. T. Dickinson, 'The Eighteenth Century Debate on the "Glorious Revolution"', *History*, vol. 28, 1976, p. 29.
60. Pincus, pp. 29–33.
61. Claydon, *Europe and the Making of England 1660–1760*, p. 242.
62. Mullett, *James II and English Politics*, pp. 68, 71–2, 81, 82.
63. Dickinson, 'The Eighteenth Century Debate on the "Glorious Revolution"', p. 32.

1 The Bishops: Unlikely Revolutionaries

1. *An Address to His Grace the Lord Archbishop of Canterbury, and the Right Reverend the Bishops Upon their Account of their Late Petition, by a true member of the Church of England*, London, 1688, p. 3.
2. Miller, *James II*, p. 31.
3. G. H. Jones, *Convergent Forces: Immediate Causes of the Revolution of 1688*, Ames IA, 1990, p. 5.
4. Bodleian Library, Oxford, Tanner MS 66, f. 180 quoted in R. A. Beddard 'William Sancroft' in *Oxford Dictionary of National Biography*, Oxford, 2005; Strickland, *The Lives of the Seven Bishops*.
5. Ibid., pp. 6–7.
6. Ibid., p. 8.
7. Ibid., pp. 8–9.
8. Bodleian Library, Tanner MS 62, f. 161.
9. G. D'Oyly, *The Life of William Sancroft, Archbishop of Canterbury*, London, 1840, p. 31.
10. Bodleian Library, Tanner MS 57, f. 525.
11. R. Bretton, 'Bishop John Lake', *Transactions of the Halifax Antiquarian Society*, 1968, pp. 89–96.
12. Carswell, *The Descent on England*, p. 136.
13. The name Thomas White was common and historians have confused him with clergy of the same name who held the offices of lectureship of St Andrew's, Holborn, and rector of St Mary-at-Hill, London.
14. W. Marshall, 'Thomas Ken', *Oxford Dictionary of National Biography*, Oxford, 2004.
15. W. Sancroft, *A sermon preached in S. Peter's Westminster, on the first Sunday in Advent, at the consecration of the Right Reverend Fathers in God, John Lord Bishop of Durham, William Lord Bishop of S. David's, Beniamin L. Bishop of Peterborough, Hugh Lord Bishop of Landaff, Richard Lord Bishop of Carlisle, Brian Lord Bishop of Chester, and John Lord Bishop of Exceter*, London, 1660.

16. R. Beddard, 'William Sancroft', *Oxford Dictionary of National Biography*, Oxford, 2004. Cosin wanted Sancroft to marry, and even found a respectable gentlewoman for him, but Sancroft's reply was that he was determined to remain single. Beddard claimed he was a celibate and his household was managed by his sister Catherine. The 'golden rectory' was so called because it had one of the largest incomes in the country.
17. D'Oyly, *The Life of William Sancroft,*, p. 276.
18. *Lex ignea, or, The school of righteousness a sermon preach'd before the King, Octob. 10. 1666 at the solemn fast appointed for the late fire in London by William Sancroft*, London, 1666. The King ordered the sermon to be published.
19. R. Beddard, 'William Sancroft', *ODNB*.
20. H. H. Poole, 'John Lake', *ODNB*. The mob forced themselves into the minster wearing their hats; Lake, who never lacked courage, pushed the hats off those who were within reach and told them either to stay and join in the service, or leave. Awed momentarily the people retired, but soon after the crowd burst open the south door and defied Lake in an attempt to provoke him. Lake, however, kept his temper, even when they followed him home, and threatened to plunder and pull down his house. Lake similarly forced the abandonment of singing in the cathedral, which sometimes led to ribaldry.
21. M. Mullett, 'Thomas White', *ODNB*.
22. P. Hopkins, 'Francis Turner', *ODNB*.
23. A. Marvel, *Mr Smirke, or the Divine in Mode...*, London, 1676, pp. 3–4.
24. Bodleian Library, MS Rawl. letters 98–99.
25. J. Spurr, *The Restoration Church of England 1646–1689*, London and New Haven CT, 1991, p. 81.
26. M. Mullett, 'William Lloyd', *ODNB*.
27. A. M. Coleby, 'Jonathan Trelawny', *ODNB*.
28. Quoted in Strickland, *The Lives of the Seven Bishops*, p. 32.
29. Ibid., p. 32. When Sancroft claimed poverty and his inability to pay the first fruits of the see of Canterbury, let alone the costs of a coach and all the equipage of the archbishop, the King brushed these aside, promising to remit the first fruits and give him funds for his day-to-day costs.
30. Strickland claims this happened at Westminster Abbey, Beddard claims it was in Lambeth Palace. William Stubbs indicates Westminster, W. Stubbs, *Registrum Sacrum Anglicanum*, Oxford, 1858.
31. Quoted in Strickland, *The Lives of the Seven Bishops*, p. 114.
32. Ibid., pp. 116–7.
33. Ibid., pp. 137–8.
34. Ibid., p. 248.
35. Plumptre, vol. 1, p. 159.
36. Quoted in Strickland, *The Lives of the Seven Bishops*, pp. 249, 256.
37. It was usual for newly consecrated bishops to give a dinner to the nobility, privy councillors and clergy; but Ken, following the example of Dr John Fell, Bishop of Oxford, devoted the sum it would have cost to the fund for the rebuilding of St Paul's cathedral. The fact is thus recorded in Dugdale's *History of St. Paul's Cathedral*, January 26. Among the list of contributors, Dr Thomas Ken, Lord Bishop of Bath and Wells, in lieu of his consecration dinner and gloves, 100l.

38. H. A. L. Rice, *Thomas Ken, Bishop and Nonjuror*, London, 1958, p. 79.
39. J. H. Overton, *The Nonjurors: Their Lives, Principles, and Writings*, London, 1902, pp. 38–46.
40. Quoted in Strickland, *The Lives of the Seven Bishops*, p. 370.
41. Smith, *Fighting Joshua*, p. 89.
42. Quoted in Strickland, *The Lives of the Seven Bishops*, pp. 373–4.
43. Ibid., pp. 34–6.
44. D'Oyly, *The Life of William Sancroft*, pp. 100–21.
45. W. Gibson, 'Dissenters, Anglicans and the Glorious Revolution: The Collection of Cases' in *The Seventeenth Century*, vol. 22, no. 1, 2007.
46. W. Gibson, 'The Limits of the Confessional State: Electoral Religion in the Reign of Charles II' in *Historical Journal*, 2008.
47. Strickland, *The Lives of the Seven Bishops*, p. 48.
48. Dillon, *The Last Revolution: 1688 and the Creation of the Modern World*, p. 5.
49. Bodleian Library, Tanner Ms 32, f. 214.
50. Beddard 'William Sancroft', *ODNB*.
51. Quoted in Strickland, *The Lives of the Seven Bishops*, p. 55.
52. Chapter 13 v. 1.
53. J. Dryden, *Astraea Redux*, London, 1660, lines 79–80. He was similarly depicted in *Absolom and Achitophel* (1681). I am grateful to Vicky Bancroft for drawing these to my attention.
54. As Shakespeare claimed in *Richard II*,
    Not all the water in the rough rude sea
    Can wash the balm from an anointed king.
    (Act 3. Scene 2).
55. 1 Kings 1 verses 32–48. I owe much of this to conversations with Dr John Morgan-Guy.
56. Quoted in Strickland, *The Lives of the Seven Bishops*, p. 124.
57. Ibid., pp. 126–7.
58. H. H. Poole, 'John Lake'.
59. Quoted in Strickland, *The Lives of the Seven Bishops*, p. 263. Ken later recalled of the Monmouth rebellion that
    in King James's time there were about a thousand or more imprisoned in my diocese who had been engaged in the rebellion of the Duke of Monmouth, and many of them were such as I had reason to believe were ill men and devoid of all religion, and yet for all that I thought it my duty to relieve them. It is well known in the diocese that I visited them night and day.
60. Quoted in Strickland, *The Lives of the Seven Bishops*, p. 261.
61. *The Diary of Dr Thomas Cartwright, Bishop of Chester*, Camden Society, 1843, p. 74. The gift of curing scrofula had been claimed by sovereigns from the days of Edward the Confessor; it was exercised after the Reformation by Queen Elizabeth, and in some of the old prayer-books a liturgy for such a service was still printed. James, however, chose to have a new office prepared and published for his use.
62. Quoted in Strickland, *The Lives of the Seven Bishops*, pp. 266–7.
63. T. Ken, *A Pastoral Letter from the Bishop of Bath and Wells to His Clergy Concerning Their Behaviour during Lent*, London, 1688. For the background

in Bath and Wells, see W. Gibson, *Religion and the Enlightenment 1600–1800, Conflict and the Rise of Civic Humanism in Taunton*, Oxford, 2006.
64. Jones, *Convergent Forces*, p. 5.
65. T. Ken, *All Glory Be to God: Thomas, Unworthy Bishop of Bath and Wells...*, London, 1688, pp. 1–2.
66. Marshall, 'Thomas Ken'.
67. Quoted in Strickland, *The Lives of the Seven Bishops*, p. 376–7.
68. *Lords Hansard*, 30 June 2005, col. 356.
69. Trevelyan, *The English Revolution, 1688–1689*, p. 8.
70. E. Bohun, *The Doctrine of Non-resistance or Passive Obedience No Way Concerned in the Controversies Now Depending between Williamites and the Jacobites*, London, 1689.
71. W. Johnston, 'Revelation and the Revolution of 1688–1689' in *Historical Journal*, vol. 48, no. 2, 2005, p. 351.

2 The King's Policies 1685–7

1. Dillon, *The Last Revolution: 1688 and the Creation of the Modern World*, p. 20.
2. Mullett, *James II and English Politics*, p. 36.
3. Harris, *Revolution: The Great Crisis of the British Monarchy*, pp. 27–9. James might have been forgiven for this if he had only read such paens as: *A Congratulatory Poem Upon the Happy Arrival of His Royal Highness, James Duke of York, At London, April 8, 1682, Written by a person of quality*, London, 1682; *To His Royal Highness, at His Happy Return from Scotland*. 'Written by a Person of Quality, London ['30 May'] 1682. – I owe these references to Newton Key.
4. Miller, *James II*, p. 120.
5. T. E. S. Clarke and H. C. Foxcroft, *A Life of Gilbert Burnet... with an Introduction by C. H. Firth*, Cambridge, 1907, p. 207.
6. E. Carpenter, *The Protestant Bishop*, London, 1956, p. 79.
7. H. Horowitz, *Revolution Politicks, the Career of Daniel Finch, Second Earl of Nottingham 1647–1730*, Cambridge, 1968, p. 39.
8. A. Browning, *Thomas Osborne, Earl of Danby and Duke of Leeds, 1632–1712*, Glasgow, 1951, vol. 1, p. 371.
9. Harris, *Revolution: The Great Crisis of the British Monarchy*, p. 50.
10. Quoted in Miller, *James II*, p. 136.
11. Ibid., p. 154.
12. Jones, *Convergent Forces*, p. 8.
13. Harris, *Revolution: The Great Crisis of the British Monarchy*, p. 48.
14. J. Ellerby, *Doctrine of Passive Obedience*, London, 1685/6.
15. F. Turner, *A Sermon Preached Before their Majesties K. James II and Q. Mary at their Coronation in Westminster Abby, April 23. 1685*, London, 1685. James's problems with Turner's sermon must have paled in comparison with those of his chaplain Thomas Jones, ('sometime Domestick and Naval Chaplain to his R. Highness the Duke of York'), who wrote *Elymas the Sorcerer: Or, a Memorial Towards the Discovery Of the Bottom of this Popish-Plot, And how far his R. Higness's Directors have been Faithful to His Honour and Interest, or the Peace of the Nation. Publish'd upon occasion of a Passage in the Late Dutchess of York's Declaration for changing her Religion*. London, 1682.

16. Havighurst, 'James II and the Twelve Men in Scarlet', p. 525.
17. Harris, *Revolution: The Great Crisis of the British Monarchy*, p. 196.
18. Miller, *James II*, p. 126.
19. Browning, *Thomas Osborne, Earl of Danby*, p. 366; Miller, *James II*, p. 124.
20. Miller, *James II*, p. 137.
21. A. T. Hart, *William Lloyd 1627–1717*, London, 1952, p. 91.
22. Browning, *Thomas Osborne, Earl of Danby*, p. 372.
23. E. Carpenter, *Thomas Tenison*, London, 1948, p. 82.
24. E. M. Thompson (ed.), *The Letters of Humphrey Prideaux to John Ellis 1674–1722*, Camden Society, 1875, pp. 142–3.
25. Browning, *Thomas Osborne, Earl of Danby*, p. 373.
26. Harris, *Revolution: The Great Crisis of the British Monarchy*, p. 97.
27. Miller, *James II*, p. 146.
28. Dillon, *The Last Revolution: 1688 and the Creation of the Modern World*, p. 57.
29. Miller, *James II*, p. 144.
30. Barrow, *The Flesh is Weak*, p. 73; Anon., *The Sentence of Samuel Johnson, at the Kings-Bench-Barr at Westminster...* London, 1686. Newton Key has pointed out to me that since Johnson was author of *Julian the Apostate*, a distinctly anti-establishment tract, it is unlikely that at this point any Anglican bishops were motivated to defend him.
31. J. C. Ryle, 'James II and the Seven Bishops' in *Light from Old Times*, London, 1902.
32. Quoted in Carpenter, *The Protestant Bishop*, p. 84.
33. Browning, *Thomas Osborne, Earl of Danby*, p. 375.
34. Horowitz, *Revolution Politicks*, p. 49. – From Christmas 1687 Petre was made a full member of the Council. For Lord Nottingham this inched closer to the intolerable. Nottingham was a staunch believer in the Test Act and was appalled by Petre's admission to the Privy Council, thereafter Nottingham absented himself from Council meetings in disgust.
35. Miller, *James II*, p. 146.
36. Havighurst, 'James II and the Twelve Men in Scarlet', p. 530.
37. T. B. Howell (ed.), *Complete Collection of State Trials and Proceedings for High Treason and Other Crimes and Misdemeanours, from the Earliest Period to the Year 1783*, London, 1816, vol. 11, case 350, col. 1165–1200.
38. Harris, *Revolution: The Great Crisis of the British Monarchy*, p. 195.
39. D. Dixon, *'Godden v Hales* revisited – James II and the dispensing power' in *Journal of Legal History*, vol. 27, no. 2, 2006.
40. Havighurst, 'James II and the Twelve Men in Scarlet', pp. 522, 530–1. At the time Roger North commented of Sancroft that 'I could know his griefs by his discourse'. Quoted in Dillon, *The Last Revolution: 1688 and the Creation of the Modern World*, p. 72.
41. Carpenter, *The Protestant Bishop*, p. 86.
42. Browning, *Thomas Osborne, Earl of Danby*, p. 376.
43. M. Ashley, *The Glorious Revolution of 1688*, London, 1966, p. 64.
44. Jones, *Convergent Forces*, p. 8.
45. Harris, *Revolution: The Great Crisis of the British Monarchy*, pp. 206–7.
46. Smith, *Fighting Joshua*, p. 40.
47. E. S. de Beer, (ed.), *The Diary of John Evelyn*, Oxford, 1955, vol. iv.

48. H. Compton, *The Bishop of London's Seventh Letter of The Conference With his Clergy, held in the year 1686*, London, 1690, pp. 3, 18.
49. Harris, *Revolution: The Great Crisis of the British Monarchy*, p. 199. In all, 228 such books were published during his reign.
50. Dillon, *The Last Revolution: 1688 and the Creation of the Modern World*, pp. 73, 74.
51. Harris, *Revolution: The Great Crisis of the British Monarchy*, p. 203.
52. Huntington Library, San Marino, California, [H. L.] Hastings Papers, Religious Box 2, folder 9. Huntingdon also claimed that he rarely attended the meetings, and that, as the deliberations of the Commission were held in private, he was unable to prove that he frequently opposed the majority view of the other commissioners.
53. A. T. Hart, *The Life and Times of John Sharp, Archbishop of York*, London, 1949, p. 93.
54. H. L., Ellesmere Ms 8554, the Earl of Huntingdon's copy of proceedings against Compton.
55. Horowitz, *Revolution Politicks*, p. 44.
56. *The History of King James's Ecclesiastical Commission*, p. 19.
57. *An Exact Account of the Whole Proceedings against the Right Reverend Father in God, Henry Lord Bishop of London before the Lord Chancellor and the Other Ecclesiastical Commissioners*, London, 1688.
58. D'Oyly, *The Life of William Sancroft*, pp. 131–9.
59. Carpenter, *The Protestant Bishop*, p. 98.
60. C. E. Whitting, *Nathaniel, Lord Crewe, Bishop of Durham 1674–1721*, London, 1940, p. 149.
61. Hart, *The Life and Times of John Sharp*, p. 101.
62. Quoted in Carswell, *The Descent on England*, p. 135 fn.
63. D'Oyly, *The Life of William Sancroft*, p. 140.
64. H. C. Foxcroft, *A Supplement to Burnet's History of My Own Time, Derived from His Original Memoirs, His Autobiography, His Letters to Admiral Herbert and His Private Meditations, all Hitherto Unpublished*, Oxford, 1902, p. 213.
65. G. W. Keeton, *Lord Chancellor Jeffreys and the Stuart Cause*, London, 1965, p. 431.
66. W. A. Speck, *Reluctant Revolutionaries*, p. 128.
67. D'Oyly, *The Life of William Sancroft*, p. 148.
68. Browning, *Thomas Osborne, Earl of Danby*, p. 377.
69. A. M. Evans, 'Yorkshire and the Revolution of 1688' in *Yorkshire Archaeological Journal*, 1929, p. 263.
70. *The Diary of Dr Thomas Cartwright, Bishop of Chester*, p. 33.
71. Miller, *James II*, p. 164.
72. Jones, *Convergent Forces*, p. 75.
73. Horowitz, *Revolution Politicks*, pp. 46, 48.
74. Miller, *James II*, p. 164.
75. *The Diary of Dr Thomas Cartwright, Bishop of Chester*, p. 44.
76. N. Sykes, *William Wake*, Cambridge, 1957, vol. 1, p. 42.
77. D. Jones, *The life of James II. late King of England. Containing an account of his birth, education, religion, and enterprizes,...till his dethronement. With the various struggles made since for his restoration;...and the particulars of his death. The whole intermix'd with divers original papers*, London, 1703, p. 184.

78. James had also, to Archbishop Sancroft's fury, allowed Walker the right to print certain religious (and Catholic) books. J. Gutch, *Collectanea curiosa; or miscellaneous tracts, relating to the history and antiquities of England and Ireland, and a variety of other Subjects. Chiefly collected and now first published, from the manuscripts of Archbishop Sancroft*... Oxford, 1781, vol. 1, pp. 288–9.
79. Miller, *James II*, p. 169.
80. Jones, *The life of James II*, p. 190.
81. W. Gibson, 'The Limits of the Confessional State: Electoral Religion in the Reign of Charles II' in *Historical Journal*, 2008.
82. A. MacIntyre, 'The College, James II and the Revolution 1687–1688' in *Magdalen College and the Crown*, Oxford, 1988.
83. G. V. Bennett, 'Loyalist Oxford and the Revolution' in L. S. Sutherland and L. G. Mitchell (eds), *The History of the University of Oxford, vol 5: The Eighteenth Century*, Oxford, 1986, p. 18.
84. C. R. Beechey, 'The Defence of the Church of England in James II's Reign', Lancaster University MA thesis, 1985, p. 90.
85. *The Diary of Dr Thomas Cartwright, Bishop of Chester*, p. 48.
86. de Beer, *The Diary of John Evelyn*, p. 546.
87. Trevelyan, *The English Revolution, 1688–1689*, p. 75.
88. Carswell, *The Descent on England*, p. 137.
89. *The Dialogue*, London, 1688.
90. Miller, *James II*, p. 172.
91. Quoted in Harris, *Revolution: The Great Crisis of the British Monarchy*, p. 263.
92. H. Fishwick (ed.), *The Notebook of the Revd Thomas Jolly 1671–1693*, Chetham Society, vol. 33, 1894, p. 82.
93. *A Copy of An Address to the King by the Bishop of Oxon to be subscribed by the Clergy of His Diocese; with the Reasons for the Subscription to the Address and the reasons against it*, London, 1687, pp. 1–3.
94. Harris, *Revolution: The Great Crisis of the British Monarchy*, pp. 209, 214–8.
95. *Letter to a Dissenter*, London, 1687, p. 252.
96. Harris, *Revolution: The Great Crisis of the British Monarchy*, pp. 253–4.
97. *The Diary of Dr Thomas Cartwright, Bishop of Chester*, p. 66.
98. Quoted in R. A. Beddard, 'Observations of a London Clergyman on the Revolution of 1688–9: Being an Except from the Autobiography of Dr William Wake' in *Guildhall Miscellany*, vol. 2, 1967, p. 407.
99. J. R. Macgrath, *The Flemings in Oxford*, vol. 2, Oxford Historical Society, vol. LXII, 1913, p. 159.
100. Harris, *Revolution: The Great Crisis of the British Monarchy*, p. 208.
101. Miller, *James II*, p. 174.
102. Huntington Library, San Marino, California, Hastings Papers, Religious Box 2, folder 9. Draft defence of Huntingdon.
103. Smith, *Fighting Joshua*, p. 46.
104. Jones, *Convergent Force*, p. 7.
105. NLW, Brongyntyn Ms 2, newsletters from John Gadbury to Sir Robert Owen, 22 September 1688.
106. Whitting, *Crewe*, p. 162.
107. *To the King's Most Excellent Majesty: The Most Humble and Faithful Advice of Your Majesty's ever Dutiful Subject and Servant, the Bishop of Durham*, London, 1688.

108. Miller, *James II*, p. 171.
109. de Beer, *The Diary of John Evelyn*, p. 571.
110. Havighurst, 'James II and the Twelve Men in Scarlet', p. 533.
111. Jones, *Convergent Forces*.
112. Newton Key's thesis, ('Politics beyond Parliament: Unity and Party in the Herefordshire Region during the Restoration Period', Cornell University PhD, 1989) shows that in counties like Herefordshire, where there were some Catholics, the campaign was by no means hopeless.
113. Kenyon, *Robert Spencer, Earl of Sunderland 1641–1702*, London, 1958, p. 171.
114. M. Ashley, *The Glorious Revolution*, p. 112. Halifax believed that, however often charters were issued, James would not have obtained a Parliament that would repeal the Test Act.
115. Miller, *James II*, pp. 178–9. Even Miller concedes that the outcome of the 'three questions' campaign was a failure.
116. Dillon, *The Last Revolution: 1688 and the Creation of the Modern World*, p. 83.
117. Kenyon, 'Introduction', p. 2.
118. Quoted in Ashley, *The Glorious Revolution of 1688*, p. 68.
119. Carpenter, *The Protestant Bishop*, p. 102.
120. R. H. George, 'The Charters Granted to English Parliamentary Corporations in 1688' in *English Historical Review*, vol. LV, 1940, p. 47.
121. NLW, Coedymaen 1, Ms 75, fol. 1.
122. Foxcroft, *A Supplement to Burnet's History of My Own Time*, pp. 220, 226.
123. Even Innocent XI was aghast at these moves. – Carpenter, *The Protestant Bishop*, p. 82.
124. Hart, *The Life and Times of John Sharp*, p. 106.
125. Clarke and Foxcroft, *A Life of Gilbert Burnet*, p. 216.
126. Foxcroft, *A Supplement to Burnet's History of My Own Time*, p. 226.
127. Harris, *Revolution: The Great Crisis of the British Monarchy*, p. 115 about 40 per cent of officers in the Irish army were Catholic.
128. Carpenter, *Thomas Tenison*, pp. 60–72.
129. Barrow, *The Flesh Is Weak*, p. 75.
130. T. Tenison, *An Argument for Union...*, London, 1683, p. 19.
131. Harris, *Revolution: The Great Crisis of the British Monarchy*, p. 153.

3 The Confrontation

1. Smith, *Fighting Joshua*, p. 43.
2. Harris, *Revolution: The Great Crisis of the British Monarchy*, p. 209.
3. Quoted in Carpenter, *The Protestant Bishop*, p. 115.
4. Kenyon, *Robert Spencer, Earl of Sunderland*, p. 174.
5. Ibid., p. 168.
6. Jones, *Convergent Forces*, pp. 69–71.
7. D. Lacey, *Dissent and Parliamentary Politics in England 1661–1689: A Study in the Perpetuation and Tempering of Parliamentarianism*, New Brunswick NJ, 1969, pp. 204–5.
8. E. Green, *The March of William of Orange through Somerset with a Notice of Other Local Events in the Time of King James II*, London, 1892, p. 46.
9. Browning, *Thomas Osborne, Earl of Danby*, vol. 1, pp. 381–2.

10. Kenyon, *Robert Spencer, Earl of Sunderland*, pp. 187–8.
11. Ibid., p. 189. George, 'The Charters Granted to English Parliamentary Corporations in 1688', p. 49. In fact James sent Winchester two further charters on 15 September and 6 November 1688. Hampshire Records Office [HRO] W/A1/28–9.
12. *The Diary of Dr Thomas Cartwright, Bishop of Chester*, p. 75.
13. George, 'The Charters Granted to English Parliamentary Corporations in 1688', p. 50.
14. Ibid., p. 54.
15. *A Proclamation by the King for Suppressing and Preventing Seditious and Unlicenced Books and Pamphlets*, 10 February 1687/8.
16. *Twenty-One Conclusions Further Demonstrating the Schism of the Church of England*, London, 1688, p. 11.
17. Kenyon, *Robert Spencer, Earl of Sunderland*, p. 191.
18. Browning, *Thomas Osborne, Earl of Danby*, p. 383.
19. Clarke and Foxcroft, *A Life of Gilbert Burnet*, p. 229.
20. Ibid., p. 231.
21. Ashley, *The Glorious Revolution*, p. 115.
22. *The Diary of Dr Thomas Cartwright, Bishop of Chester*, p. 47.
23. D'Oyly, *The Life of William Sancroft*, p. 153.
24. *The History of King James's Ecclesiastical Commission*, p. 53.
25. R. Thomas, 'The Seven Bishops and Their Petition, 18 May 1688' in *The Journal of Ecclesiastical History*, vol. 12, 1961, p. 57.
26. *A Pastoral Letter from the Four Catholic Bishops to the Lay Catholics of England*, London, 1688, p. 5.
27. Miller, *James II*, pp. 166, 167.
28. M. Mullett, 'Recusants, Dissenters and North-West Politics between the Restoration and the Glorious Revolution' in J. Appleby and P. Dalton (eds), *Government, Religion and Society in Northern England 1000–1700*, Stroud, 1997, p. 208; and idem 'The Politics of Liverpool 1660–1688' in *Transactions of the Historical Society of Lancashire and Cheshire*, vol. 124, 1972.
29. Johnston, 'Revelation and the Revolution of 1688–1689', p. 360.
30. The full title of this work was, *A Collection of Cases, and other Discourses, lately written to recover Dissenters to the Communion of the Church of England, by some divines of the City of London*.
31. W. Gibson, 'Dissenters, Anglicans and the Glorious Revolution: *The Collection of Cases*'.
32. Miller, *James II*, p. 185.
33. Beddard, 'Observations of a London Clergyman on the Revolution of 1688–9'.
34. Ibid., p. 414.
35. One of these declarations was an apology for dissolving Parliament the other was his condemnation of the Rye House Plot. Hart, *William Lloyd*, p. 94. Charles had also ordered his reasons for dissolving the Parliaments of 1681 in *His Majesty's Declaration to all His Loving Subjects, Touching the Causes and Reasons That Moved him to Dissolve the Two Last Parliaments*, London, 1681.
36. Jones, *Convergent Forces*, p. 9.
37. Mullett, *James II and English Politics*, p. 68.

38. R. A. Beddard, 'A Kingdom Without a King' The Journal of the Provisional Government in the Revolution of 1688, Oxford, 1988, pp. 15–16.
39. Beechey, 'The Defence of the Church of England in James II's Reign', p. 123.
40. Jones, *Convergent Forces*, p. 122.
41. Smith, *Fighting Joshua*, p. 44.
42. Jones, *Convergent Forces*, p. 21.
43. Thomas, 'The Seven Bishops and Their Petition', p. 69.
44. G. S. de Krey, '"Reformation and Arbitrary Government": London Dissenters and James II's Polity of Toleration, 1687–1688' in J. McElligott (ed.), *Fear, Exclusion and Revolution, Roger Morrice and Britain in the 1680s*, Aldershot, 2006, p. 28.
45. *Great and Good News for the Church of England If they Please to Accept thereof: Or the Latitudinarian Christians Most Humble Address and Advice*, London, 28 May 1688. Published 'with allowance'.
46. Quoted in Harris, *Revolution: The Great Crisis of the British Monarchy*, p. 239. This happened in November 1686.
47. G. de Forest Lord et al., *Poems on Affairs of State*, New Haven CT, 1963–75, vol. 4, p. 221.
48. NLW, Brongyntyn Ms, Gadbury to Owen, 26 May 1688.
49. Jones, *Convergent Forces*, p. 9.
50. Thomas, 'The Seven Bishops and Their Petition', p. 59.
51. Gibson, 'Dissenters, Anglicans and the Glorious Revolution: *The Collection of Cases*'.
52. Jones, *Convergent Forces*, p. 11.
53. Thomas, 'The Seven Bishops and Their Petition', p. 62.
54. NLW, Brongyntyn Ms 2, Gadbury to Owen, 26 May 1688.
55. Thompson, *The Letters of Humphrey Prideaux*, p. 147.
56. Fowler had begun his career as a Presbyterian and conformed in 1662 but had retained the confidence of his Dissenting fellows. Fowler welcomed that James's policies forced Anglicans and Dissenters 'to be united in affection and to have more charity for each other'. M. Goldie and J. Spurr, 'Politics and the Restoration Parish: Edward Fowler and the Struggle for St Giles Cripplegate' in *English Historical Review*, 1994.
57. Hart, *William Lloyd*, p. 94.
58. A. Taylor, *The Works of Simon Patrick*, London, 1858, vol. 9, pp. 509–11; Carpenter, *Thomas Tenison*, p. 85.
59. Thomas, 'The Seven Bishops and Their Petition', p. 63.
60. Beechey, 'The Defence of the Church of England in James II's Reign', p. 153.
61. Thomas, 'The Seven Bishops and Their Petition', p. 60.
62. Jones, *Convergent Forces*, p. 10.
63. Ashley, *The Glorious Revolution*, p. 117.
64. Jones, *Convergent Forces*, p. 16.
65. Bishop Smith of Carlisle later wrote to Sancroft to complain that the petition was only in the name of the bishops and clergy of the province of Canterbury. This was presumably because York was vacant. J. Gutch, *Collectanea curiosa*, vol. 1, p. 334.
66. Quoted in Ashley, *The Glorious Revolution*, p. 201.
67. Horowitz, *Revolution Politicks*, p. 50.
68. NLW, Kemys-Tynte Ms, c. 124.
69. D. H. Woodforde, *Woodforde Diary and Papers*, London, 1932, p. 17.

70. *The Learned and Loyal Abraham Cowley's Definition of a Tyrant... in his Discourse concerning the Government of Oliver Cromwell*, London, 1688.
71. NLW, Brongyntyn Ms 1, Gadbury to Owen, 26 May 1688.
72. A reference to the book of Samuel in which Sheba raised a standard of rebellion against the King.
73. This part of the account of the King's reply was sent to Edward Clark, the exclusionist Whig MP for Taunton, at Chipley, HRO, 9M73/G247/6.
74. The fullest account of the interview is in D'Oyly's life of Sancroft, pp. 160–2.
75. Dryden's, *The Hind and the Panther*, quoted in M. Goldie, 'The Political Thought of the Anglican Revolution', p. 103.
76. Miller, *James II*, p. 185.
77. Anon., *What has been may be: Or a View of a Popish and an Arbitrary Government*, p. 49.
78. Carswell, *The Descent on England*, p. 140.
79. Ryle, 'James II and the Seven Bishops'.
80. Strickland, *The Lives of the Seven Bishops*, p. 63.
81. Hart, *William Lloyd*, p. 100.
82. Carswell, *The Descent on England*, p. 139. Carswell claims that the printer, Clavell in St Paul's churchyard, was often used by Lloyd.
83. NLW, Coedymaen Ms, 1, ff. 41, 42, 43, 44. fol. 41 is endorsed 'after the petition was presented'.
84. J. Gutch, *Collectanea curiosa*, vol. 1, p. 373.
85. Carpenter, *The Protestant Bishop*, p. 117. See for example Horsford, 'Bishop Compton and the Revolution on 1688', p. 210. The same points can be made for Sancroft who was at the Lambeth meeting but did not go to present the petition to the King, but no one has suggested that he might have been responsible for so cynical an act as the leak of the petition.
86. Jones, *Convergent Forces*, p. 123.
87. Ibid., p. 15.
88. Thomas, 'The Seven Bishops and Their Petition', p. 67.
89. Ibid., p. 67.
90. Carswell, *The Descent on England*, p. 143.
91. *An Answer to a Paper importing a Petition of the Archbishop of Canterbury and Six other Bishops, to His Majesty, touching their not Distributing and Publishing The Late Declaration for Liberty of Conscience*, London, 1688.
92. Miller, *James II*, p. 186.
93. *The Examination of the Bishops Upon their Refusal of Reading His Majesty's Most Gracious Declaration; And the Nonconcurrence of the Church of England, In Repeal of the Penal Laws and Test Fully Debated and Argued*, London, 1688, pp. 1, 10–11, 12–13, 21.
94. Ibid., pp. 23, 28, 32, 33–8.
95. Its full title was: *Toleration Tolerated, or A Late Learned Bishop's Opinion Concerning Toleration of Religion*, London, 1688.
96. *Toleration Tolerated, Or a Late Learned Bishop's Opinion Concerning Toleration of Religion.*
97. The full title being: *An Address to His Grace the Lord Archbishop of Canterbury, and the Right Reverend the Bishops Upon their Account of their Late Petition, by a true member of the Church of England*, London, 1688.
98. Ibid., p. 3.

99. Ibid., pp. 6–12.
100. *Some Queries Humbly Offered to the Lord Archbishop of Canterbury and the Six Other Bishops concerning the English Reformation And the Thirty Nine Articles of the Church of England*, London, 1688.
101. Carswell, *The Descent on England*, p. 138.

4 The Tower

1. de Beer, *The Diary of John Evelyn*, p. 584.
2. NLW, Brongyntyn Ms 1, Gadbury to Own, 26 May 1688.
3. Hart, *William Lloyd*, p. 101; Taylor, *Works of Simon Patrick*, p. 512.
4. HRO, 9M73/G247/6.
5. Carswell, *The Descent on England*, p. 140.
6. de Beer, *The Diary of John Evelyn*, p. 585.
7. Hart, *William Lloyd*, p. 101.
8. *Public Occurrences*, 12 June 1689, quoted in Thomas, 'The Seven Bishops and Their Petition', p. 68; E. Green, *The March of William of Orange through Somerset*.
9. NLW, Coedymaen Ms 1, 42.
10. Carswell, *The Descent on England*, p. 142–8.
11. NLW, Coedymaen Ms 1, f. 39.
12. Kenyon, *Robert Spencer, Earl of Sunderland*, p. 196.
13. Jones, *The life of James II*, p. 179.
14. *The Convocation*, London, 1688, Brotherton Collection of Manuscript Verse, Leeds University, item: Crum C72; POAS V. 130–6.
15. *A Letter from a Clergyman in the City to his friend in the country, Containing his Reasons For not Reading the Declaration*, London, 1688. See also Jones, *Convergent Forces*, p. 17.
16. *The Countrey-Minister's Reflections on the City Minister's Letter to his Friend, Shewing the Reasons why We cannot Read the King's Declaration in Our Churches*, London, 1688.
17. *The Minister's Reasons For His Not reading the King's Declaration, Friendly Debated by a Dissenter*, London, 1688.
18. *A Letter of several French Ministers Fled into Germany upon the account of the Persecution in France, to such of their Brethren in England as Approved the King's Declaration touching Liberty of Conscience, translated from the Original in French*, London, 1688, p. 2.
19. Ibid., pp. 3–7.
20. Taylor, *Life of Patrick*, p. 512.
21. Harris, *Revolution: The Great Crisis of the British Monarchy*, p. 261. Harris claims no one in Oxford read the Declaration.
22. C. A. Lowe, 'Politics and Religion in Warwickshire during the Reign of James II, 1685–1688', Warwick University MA thesis, 1992, p. 146.
23. Jones, *Convergent Forces*, pp. 18–19.
24. Gutch, *Collectanea curiosa*, vol. 1, pp. 331–2. Beaw was, like Mews, an old Cavalier bishop who had fought for Charles I and had been a secret agent for the exiled Stuarts. J. R. Guy, 'William Beaw: Bishop and Secret Agent', *History Today*, vol. 26, no. 12, 1976, pp. 796–803.

25. N. Key, 'Comprehension and the Breakdown of Consensus in Restoration Herefordshire' in T. Harris, P. Seaward and M. Goldie (eds), *The Politics of Religion in Restoration England*, Oxford, 1990, p. 207. Croft had suppressed the Jesuit College at Cwm. – I owe this to John Morgan-Guy.
26. British Library, Add. Mss, 34, 510, f. 133.
27. Ryle, 'James II and the Seven Bishops'.
28. Whitting, *Crewe*, p. 177.
29. M. Storey, *Two East Anglian Diaries, 1641–1729*, Suffolk Records Society, vol. XXXVI, 1994, p. 175.
30. G. V. Bennett, *White Kennett, 1660–1728*, London, 1957, p. 11.
31. Jones, *Convergent Forces*, pp. 9, 18.
32. Ibid., p. 19.
33. Heitzman, 'The Revolution of 1688 in Lancashire and Cheshire', p. 81. In August 1688 he even presented an address from his clergy expressing their thanks for the Declaration and condemning the seven bishops.
34. *HMC Buccleugh*, ii:32.
35. Bodleian Library, Tanner, 29, f. 12. He was active in promoting the first Declaration, issued shortly before his own elevation, writing to Bishop Lloyd of Norwich in April 1687 in a vein that suggested all the bishops supported it and asking him to secure a supportive address to the King from the clergy of Norfolk and Suffolk. Watson was unequivocal about his own attitude to such an address: 'I can see no harm or ill consequence in the thing I think it a fair declaration of ... confidence in his Majesty's protection, and a better method than fears and jealousies.'
36. Lambeth Palace Library, VX 1B 2g/2, box 1, deposition of Hugo Powell.
37. NLW, Brongyntyn Ms 1, Gadbury to Owen, 9 June 1688.
38. E. Ellis, *A Clergyman of the Church of England: His Vindication of Himself for Reading His Majesties Late Declaration*, London, 1688.
39. Quoted in Jones, *Convergent Forces*, pp. 9–10.
40. *A Dialogue between the ArchB of C. and the Bishop of Heref. Containing the True reasons why the Bishops could not read the Declaration*, London, 1688.
41. Jones, *Convergent Forces*, p. 16.
42. J. S. Clarke, *The Life of James II by Himself*, London, 1816, vol. 2, pp. 155–6.
43. A. Browning (ed.), *Memoirs of Sir John Reresby*, London 1991 (second edition with notes by M. Geiter and W. Speck), p. 499.
44. Keeton, *Lord Chancellor Jeffreys and the Stuart Cause*, p. 434.
45. The Privy Council members present were the Lord Chancellor, the Lord Privy Seal, Lord President, Lords Powis, Huntingdon, Peterborough, Craven, Berkeley, Moray, Middleton, Melfort, Castlemaine, Preston, Dartmouth, Godolphin, Dover, the Lord Chief Justice Herbert, Sir Nicholas Butler and Fr Petre. NLW Coedymaen Ms, 1, 45.
46. Jones, *Convergent Forces*, p. 21.
47. This account of the second interview with James is from D'Oyly, pp. 167–70.
48. The warrant was signed by Lord Chancellor Jefferies, Lords Sunderland, Arundel, Powis, Mulgrave, Huntingdon, Peterborough, Craven, Murray, Middleton, Melfort, Castlemain, Darmouth, Godolphin, Dover, and Sir John Ernle, Sir Edward Herbert and Sir Nicholas Butler. *A Collection of the Most Remarkable and Interesting Trials. Particularly of those Persons who have forfeited*

their lives to the injured laws of their country. In which the most remarkable of the State Trials will be included, London, 1775–6, vol. 1, p. 714.
49. de Beer, *The Diary of John Evelyn*, p. 586.
50. Lancashire Record Office, Kenyon of Peel Mss, DDKE/acc. 7840.
51. Jones, *Convergent Forces*, p. 21.
52. *The Confinement of the Seven Bishops*, London, 1688.
53. Hart, *William Lloyd*, p. 107.
54. Ryle, 'James II and the Seven Bishops'.
55. Quoted in C. Knight, *History of England*, London, 1864, vol. iv, p. 419.
56. Barone, *Our First Revolution*, p. 132.
57. O. Heywood, *Works*, 1882–5, vol. I, p. 287.
58. *Memoirs and Travels of Sir John Reresby*, London, 1904, p. 302.
59. Lancashire Record Office, Kenyon of Peel Mss, DDKE/acc. 7840 Lancashire Trials November 23–5 February 1694/5.
60. Carpenter, *The Protestant Bishop*, p. 121.
61. Barone, p. 131.
62. Strickland, *The Lives of the Seven Bishops*, pp. 67–8.

5 The Trial

1. W. L. Sachse, *Lord Somers, A Political Portrait*, Manchester, 1975, 21. Somers was subsequently to be the author of *A brief history of the succession of the crown of England, &c. collected out of the records, and the most authentick historians, written for the satisfaction of the nations*, London, 1689, which argued that the succession was a matter to be determined by Parliament.
2. Jones, *Convergent Forces*, p. 29.
3. H. Howitz (ed.), *The Parliamentary Diary of Narcissus Luttrell, 1691–1693*, Oxford, 1972, vol. 1, p. 448.
4. NLW, Coedymaen Ms 1, 46. Among the witnesses that the King had received the petition on 20 May was Samuel Pepys.
5. Harris, *The Politics of Religion in Restoration England*, p. 265.
6. Carswell, *The Descent on England*, pp. 143, 144.
7. *An Account of the Proceedings at the King's Bench Bar at Westminster Hall Against the Seven Bishops*, London, 16 June 1688.
8. Browning, *Thomas Osborne, Earl of Danby*, p. 385.
9. de Beer, *The Diary of John Evelyn*, p. 587.
10. The extracts, particularly of the arraignment and trial in this chapter, where not otherwise referenced are from: *The Proceedings and Tryal in the Case of the most Reverend Father in God, William, Lord Archbishop of Canterbury and the Right Reverend Fathers in God, William, Lord Bishop of St. Asaph, Francis, Lord Bishop of Ely, John, Lord Bishop of Winchester, Thomas, Lord Bishop of Bath and Wells, Thomas, Lord Bishop of Peterborough, and Jonathan, Lord Bishop of Bristol, in the court of Kings-Bench at Westminster in Trinity-term in the fourth year of the reign of King James the Second, Annoque Dom. 1688*, London, 1689 and *A Collection of the Most Remarkable and Interesting Trials. Particularly of those Persons who have forfeited their lives to the injured laws of their country. In which the most remarkable of the State Trials will be included*, London, 1775–6, vol. 1, pp. 712–48.

Notes   221

11. Carswell, *The Descent on England*, p. 147.
12. W. Gibson, '"Look toward the Court!" Two Jesus Men in 1688' in *Jesus College Record*, 2008.
13. Jones, *Convergent Forces*, p. 28.
14. *Trinity Term, 4 Jac. 2, in B. R.*
15. Ryle, 'James II and the Seven Bishops.'
16. R. A. Beddard, 'Two Letters from the Tower 1688' in *Notes and Queries*, September 1984, pp. 347–52.
17. Hart, *William Lloyd*, p. 109.
18. Jones, *Convergent Forces*, p. 29.
19. Beechey, 'The Defence of the Church of England in James II's Reign', p. 156.
20. Carswell, *The Descent on England*, p. 143.
21. Keeton, *Lord Chancellor Jeffreys and the Stuart Cause*, p. 406.
22. The jury was comprised of Sir Roger Langley, Sir William Hill, Roger Jennings, Thomas Harriot, Jeophery Nitingall, Will Withers, Will Avery, Thomas Austin, Nicholas Grace, Michael Arnold, Thomas Done, Richard Shoreditch. – Jones, *The life of James II*, p. 194.
23. Jones, *Convergent Forces*, p. 31.
24. W. J. Smith, *The Herbert Correspondence: The Sixteenth and Seventeenth Century Letters of the Herberts of Chirbury*, Cardiff, 1968, pp. 339–40.
25. Beinecke Library, Yale University, Osborn Collection, William Bagot letters, FB190.
26. *Letter from a Country Curate to Mr Henry Care in Defence of the Seven Bishops...*, London, 18 June 1688.
27. Sykes, *William Wake*, p. 43.
28. 'Letter of Bishop Levinze to Thomas Choldmondley', *Proceedings of the Manx Society*, vols IV, VII and IX.
29. *Two Plain Words to the Clergy: Or an Admonition to Peace and Concord at this Juncture*, London, 1688.
30. Carpenter, *The Protestant Bishop*, p. 121.
31. NLW, Plas Yn Cefn Ms 2939.
32. Keeton argued that Jeffreys sought to restrain James and that even Clarendon conceded that Jeffreys exerted a moderating influence on the King. – Keeton, *Lord Chancellor Jeffreys and the Stuart Cause*, p. 405.
33. Miller, *James II*, p. 187.
34. Jones, *Convergent Forces*, p. 151.
35. Strickland, *The Lives of the Seven Bishops*, pp. 350–1.
36. *A Poem on the Deponents Concerning the Birth of the Prince of Wales*, London, 1688, p. 4.
37. *The Sham Prince Expos'd In A Dialogue between the Pope's Nuncio and Bricklayer's Wife, Nurse to the Supposed Prince of Wales*, London, 1688.
38. NLW, Brogyntyn Ms 1, Gadbury to Owen, 9 June 1688.
39. Havighurst, 'James II and the Twelve Men in Scarlet', p. 540. 'Halter' meant noose.
40. Jones, 'The Invitation to William of Orange in 1688', p. 33.
41. Who had been persuaded by Tenison to make 'a splendid appearance', Jones, *Convergent Forces*, p. 31.
42. Browning, *Memoirs of Sir John Reresby*, p. 501.
43. Jones, *Convergent Forces*, pp. 36, 37.

44. Carswell, *The Descent on England*, p. 148.
45. Horowitz, *Revolution Politicks*, p. 51.
46. W. L. Sachse, 'The Mob and the Revolution of 1688' in *The Journal of British Studies*, vol. 4, 1964, p. 22.
47. And in 1685 after the Monmouth emergency the Commons had told the King that 'those [Catholic] officers could not by law be capable of their employments; and that the incapacities they bring upon themselves thereby, can no ways be taken off, but by Act of Parliament'. *A Collection of the Most Remarkable and Interesting Trials. Particularly of those Persons who have forfeited their lives to the injured laws of their country. In which the most remarkable of the State Trials will be included*, London, 1775–6, vol. 1, p. 733.
48. Speck, *Reluctant Revolutionaries*, p. 152.
49. Quoted in Havighurst, 'James II and the Twelve Men in Scarlet', p. 545.
50. Hart, *William Lloyd*, p. 111.
51. Jones, *Convergent Forces*, p. 41.
52. Dillon, *The Last Revolution: 1688 and the Creation of the Modern World*, p. 121.
53. Jones, *Convergent Forces*, p. 30.
54. Gutch, *Collectanea curiosa*, vol. 1, p. 374.
55. HRO, 9M73/G239/6 Edward Clark to Thomas Stringer, 30 June 1688. Clark wrote of the trial that the bishops and lawyers behaved themselves with 'great courage and conduct and spoke such Bold Truths concerning the Dispensing Power as have not of latter years been mentioned in Westminster Hall'.
56. Anon., *What has been may be: Or a View of a Popish and an Arbitrary Government*, p. 69.
57. Dillon, *The Last Revolution: 1688 and the Creation of the Modern World*, p. 121.
58. Browning, *Memoirs of Sir John Reresby*, p. 501.
59. Jones, *Convergent Forces*, p. 42.
60. Carswell, *The Descent on England*, p. 149.
61. NLW, Kemys-Tynte Ms, c. 128.
62. *The History of King James's Ecclesiastical Commission*, p. 61.
63. London Metropolitan Archives, Middlesex Sessions of the Peace, MJ/SP/1692, file 3.
64. Jones, *Convergent Forces*, pp. 42–3.
65. Sachse, 'The Mob and the Revolution of 1688', p. 25.
66. Johnston, 'Revelation and the Revolution of 1688–1689', p. 363.
67. *The Prince of Orange Welcome to London*, London, 1688. Pepys Library, II, 255. I owe this and other broadside ballads to Dr Angela McShane-Jones.
68. *A Third Touch of the Times*, Pepys IV, 311.
69. *A View of the Popish Plot*, Pepys II, 281.
70. Robert H. Murray (ed.), *The Journal of John Stevens Containing a Brief Account of the War in Ireland 1689–1691*, Oxford, 1912, pp. 4–5.
71. Harris, *The Politics of Religion in Restoration England*, p. 268.
72. Taylor, *Works of Simon Patrick*, p. 513.
73. Dillon, *The Last Revolution: 1688 and the Creation of the Modern World*, p. 117.
74. Gutch, *Collectanea curiosa*, vol. 1, p. 384.
75. Harris, *The Politics of Religion in Restoration England*, p. 268.
76. Miller, *James II*, p. 188.
77. Browning, *Memoirs of Sir John Reresby*, p. 502.
78. Macgrath, *The Flemings in Oxford*, p. 220n.

79. Harris, *The Politics of Religion in Restoration England*, pp. 228, 270.
80. Jones, *Convergent Forces*, p. 43.
81. S. W. Rix (ed.), *The Diary and Autobiography of Edmund Bohun*, Beccles, 1853, p. 81; Bohun dared not to publicly question the birth of the Prince, but he noticed that women sniggered when prayers were said for him in church.
82. Storey, *Two East Anglian Diaries*, p. 176.
83. de Krey, 'Reformation and Arbitrary Government', p. 29.
84. M. Goldie, *Roger Morrice and the Puritan Whigs, The Entering Book of Roger Morrice, 1677–1691*, Woodbridge, 2007, pp. 52–3.
85. Lowe, 'Politics and Religion in Warwickshire', pp. 147, 152, 160.
86. Sachse, 'The Mob and the Revolution of 1688', p. 25.
87. Bristol Record Office, P/StJ/ChW/1b.
88. Trelawny's home was, of course, within Exeter diocese at that time.
89. J. Barry, 'Exeter in 1688: The Trial of the Seven Bishops' in *Devon and Cornwall Notes and Queries*, vol. 38, 1996.
90. Smith, *Fighting Joshua*, p. 49.
91. Dillon, *The Last Revolution: 1688 and the Creation of the Modern World*, p. 125.
92. Evelyn Cruickshanks has shown that there were some loyal displays for James, mainly in the North and East. E. Cruickshanks, 'The Revolution and the Localities: Examples of Loyalty to James II' in E. Cruickshanks (ed.), *By Force or by Default? The Revolution of 1688–1689*, Edinburgh, 1989, pp. 28–41.
93. *The Clergy's Late Carriage to the King Considered in a Letter to a Friend*, London, 1688, pp. 1, 2.
94. Jones, *Convergent Forces*, p. 43.
95. Carswell, *The Descent on England*, p. 140 fn.
96. Gutch, *Collectanea curiosa*, vol. 1, p. 384. Mrs Clarke may have been Mary, wife of Edward Clark, MP for Taunton.
97. Trevelyan, *The English Revolution, 1688–1689*, p. 96.

6 The Reaction

1. Havighurst, 'James II and the Twelve Men in Scarlet', pp. 539–40.
2. Sachse, 'The Mob and the Revolution of 1688', p. 23.
3. R. A. Beddard, 'The Unexpected Whig Revolution of 1688' in Beddard (ed.), *The Revolutions of 1688*, Oxford, 1991, p. 99.
4. T. Harris, 'London Crowds and the Revolution of 1688' in E. Cruickshanks (ed.), *By Force or by Default? The Revolution of 1688–1689*, Edinburgh, 1989, pp. 44–51.
5. Evans, 'Yorkshire and the Revolution of 1688', p. 266.
6. Keeton, *Lord Chancellor Jeffreys and the Stuart Cause*, p. 444.
7. Carswell, *The Descent on England*, p. 150 fn.
8. Cipher for Russell.
9. Carpenter, *The Protestant Bishop*, p. 123.
10. Miller, *James II*, p. 187.
11. Carswell, *The Descent on England*, p. 151.
12. Harris, 'London Crowds and the Revolution of 1688', p. 268.
13. Horowitz, *Revolution Politicks*, p. 52.
14. Dillon, *The Last Revolution: 1688 and the Creation of the Modern World*, p. 125.

15. Carpenter, *The Protestant Bishop*, p. 123.
16. Taylor, *Works of Simon Patrick*.
17. Clarke and Foxcroft, *A Life of Gilbert Burnet*, p. 241.
18. Quoted in Jones, *Convergent Forces*, p. 133.
19. This was the origin of the non-juror position.
20. *The Autobiography of Symon Patrick*, London, 1839, pp. 138–9.
21. NLW, Brongyntyn Ms 1, Gadbury to Owen, 22 September 1688.
22. D. Davies, 'James II, William of Orange and the Admirals' in E. Cruickshanks (ed.), *By Force or by Default?*, p. 90.
23. Kenyon, *Robert Spencer, Earl of Sunderland*, pp. 207, 208.
24. D. L. Jones, 'The Glorious Revolution in Wales' in *The National Library of Wales Journal*, vol. 26, no. 1, 1989, pp. 27–31.
25. 'dooms' meaning 'judges'.
26. W. Thomas, *The Mammon of Unrighteousness Detected and Purified in a Sermon Preached in the Cathedral Church of Worcester on Sunday the Nineteenth of August, 1688*, London, 1688, pp. 3, 12, 32.
27. Dorset Record Office, Poole Borough Papers, DC/PL/A/1/16. In a last attempt to win back the borough in December, James restored the original borough charter, but it was too late.
28. NLW, Brongyntyn Ms. PNQ 3/1/20.
29. Havighurst, 'James II and the Twelve Men in Scarlet', p. 541.
30. Jones, *Convergent Forces*, p. 45. John Gadbury, a strong supporter of James, wrote to Sir Robert Owen on 19 July that there were 'grumblings' amid 'hot headed clergymen' that they were going to be prosecuted for refusal to read the Declaration 'but some of them are come to themselves and own the King's prerogative to be above them'. He also commented that he expected some 'ecclesiastical jarrings' for 'love makes single persons mad but religion makes whole nations delirious'. – NLW Brongyntyn 1, Gadbury to Owen, 19 July 1688.
31. Jones, *The life of James II*, p. 195.
32. Jones, *Convergent Forces*, pp. 46–9.
33. *The Articles Recommended by the Archbishop of Canterbury to all Bishops within his Metropolitan Jurisdiction, the 16th of July, 1688*, London, 1688, p. 1.
34. In Birmingham and Worcestershire there were 176 converts. – Speck, *Reluctant Revolutionaries*, p. 182.
35. *The Articles Recommended by the Archbishop of Canterbury to all Bishops within his Metropolitan Jurisdiction, the 16th of July, 1688*, London, 1688.
36. de Beer, *The Diary of John Evelyn*, p. 596. 'To the Right Honourable My Lords of his Majesty's Commission Ecclesiastical' in *A Fourth Collection of Papers relating to the Present Juncture of Affairs in England*, London, 1688, p. 17. The Bishop of Rochester's letter refusing to attend the Commission was quickly made public.
37. Huntington Library, San Marino, California, Hastings Papers, Religious Box 2, folder 9.
38. Smith, *Fighting Joshua*, p. 49.
39. NLW, Brongyntyn Ms PQN 3/1/23–5.
40. Quoted in Ryle, 'James II and the Seven Bishops'.
41. M. Mendle, 'The "prints" of the Trials: The Nexus of Politics, Religion, Law and Information in late seventeenth century England' in J. McElligott (ed.), *Fear, Exclusion and Revolution*, p. 123.

42. Spurr, *The Restoration Church of England*, p. 97
43. Sachse, 'The Mob and the Revolution of 1688', pp. 27–8, 30.
44. Barone, *Our First Revolution*, p. 135.
45. Havighurst, 'James II and the Twelve Men in Scarlet', pp. 540–1. NLW Brongyntyn Ms, PQN 3/1/21–2.
46. Havighurst, 'James II and the Twelve Men in Scarlet', p. 542.
47. NLW, Brongyntyn Ms, PQN 3/1/17–20.
48. Jones, *Convergent Forces*, p. 50.
49. A. Coleby, *Central Government and the Localities: Hampshire 1649–1689*, Cambridge, 1987, p. 205.
50. Lowe, 'Politics and Religion in Warwickshire', p. 128.
51. Havighurst, 'James II and the Twelve Men in Scarlet', p. 542.
52. For example see 'A Memorial of the Protestants of the Church of England to their Royal Highnesses the Prince and Princess of Orange' in *A Collection of Papers relating to the Present Juncture of Affairs in England*, London, 1688, p. 30.
53. F. P. Verney and M. M. Verney (eds), *Memoirs of the Verney Family, 1642–96*, London, 1892, vol. 1, p. 458.
54. Havighurst, 'James II and the Twelve Men in Scarlet', p. 542.
55. Bodleian Library, MS Tanner 28, fol. 178.
56. *A Prophylactick from Disloyalty in these Perilous Times. In a Letter to the Right Honourable and Right Reverend Father in God Herbert, By Divine Providence, Lord Bishop of Hereford*, London, 1688.
57. *Nahash Revived: Or the Church of England's Love to Dissenters and Loyalty to their Prince*... London, 1688, pp. 1, 3, 4, 5, 6.
58. Green, *The March of William of Orange through Somerset*, p. 48.
59. Beddard, 'Observations of a London Clergyman on the Revolution', p. 410.
60. Ibid., p. 414.
61. Calamy quoted in H. S. Skeats *A History of the Free Churches of England*, London 1867, p. 90.
62. G. Burnet, *Apology for the Church of England with relation to the Spirit of Persecution*, London, 1688, p. 6.
63. de Beer, *The Diary of John Evelyn*, p. 590.
64. Kenyon, *Robert Spencer, Earl of Sunderland*, p. 201.
65. Derbyshire Record Office, Gell Papers, D258/17/31/58.
66. *Ballad to the Tune of Couragio*, London, 1688.
67. Rix, *The Diary and Autobiography of Edmund Bohun*, p. 81.
68. Jones, *Convergent Forces*, p. 169.
69. Miller, *James II*, pp. 196–7.
70. At Christmas Morrice claimed that Archbishop was 'politically sick'. Dillon, *The Last Revolution: 1688 and the Creation of the Modern World*, p. 211.
71. Dillon, *The Last Revolution: 1688 and the Creation of the Modern World*, p. 143.
72. Jones, *The life of James II*, p. 199.
73. 'Some Account of the Humble Application of the Pious and Noble Prelate, Henry Lord Bishop of London, with the Reverend Clergy of the City and some of the Dissenting Ministers in it, to the Illustrious Prince William Henry...' in *A Sixth Collection of Papers relating to the Present Juncture of Affairs in England*, London, 1688, p. 17.
74. NLW, Brongyntyn Ms, PQN 1/3/31.
75. Carpenter, *The Protestant Bishop*, pp. 124–5.
76. de Beer, *The Diary of John Evelyn*, p. 597.

77. Jones, *Convergent Forces*, pp. 135–6.
78. Dillon, *The Last Revolution: 1688 and the Creation of the Modern World*, p. 144.
79. These items were widely published in 'An Account of the Late Proposals of the Archbishop of Canterbury, with some Other Bishops to his Majesty: In a Letter to M. B.' in *A Collection of Papers relating to the Present Juncture of Affairs in England*, London, 1688, pp. 7–10. Jones, *Convergent Forces*, pp. 138–9.
80. Miller, *James II*, p. 198.
81. Gutch, *Collectanea curiosa*, vol. 1, p. 420.
82. Harris, 'London Crowds and the Revolution of 1688', p. 278.
83. *The History of King James's Ecclesiastical Commission*, p. 54.
84. Lancashire Record Office, Kenyon of Peel Mss, DDKE/acc. 7840, 28 Sept–21 Nov 1688.
85. Heitzman, 'The Revolution of 1688 in Lancashire and Cheshire', p. 112.
86. Anon., *An Account of the Proposals of the Archbishop of Canterbury with some other Bishops to His Majesty, In a Letter to M. B. Esq.*, London, 1688.
87. Macgrath, *The Flemings in Oxford*, pp. 220, 233.
88. Jones, *Convergent Forces*, p. 138.
89. de Beer, *The Diary of John Evelyn*, p. 600.
90. Jones, *Convergent Forces*, p. 140; Miller, *James II*, p. 196.
91. Whiting, *Crewe*, p. 181.
92. Miller, *James II*, p. 198.
93. Jones, *Convergent Forces*, p. 148.
94. Huntington Library, San Marino, California, 'Memorandum for those that go into the Country to dispose the Corporations to a good Election for Members of Parliament. To be read by them often'. October 1688.
95. Gutch, *Collectanea curiosa*, vol. 1, p. 419.
96. de Beer, *The Diary of John Evelyn*, p. 601.
97. Jones, *The life of James II*, p. 202.
98. NLW, Brongyntyn Ms, PQN 1/3/31.
99. Jones, *Convergent Forces*, p. 141.
100. Horowitz, *Revolution Politicks*, p. 55.
101. 'The Prince of Orange's first Declaration from the Hague...with his Highnesses Additional Declaration from the Hague, October 24 1688' in *A Fourth Collection of Papers relating to the Present Juncture of Affairs in England*, London, 1688, p. 5.
102. de Beer, *The Diary of John Evelyn*, p. 602.
103. Jones, *Convergent Forces*, p. 145.
104. Quoted in Dillon, *The Last Revolution: 1688 and the Creation of the Modern World*, pp. 160–1.
105. Horowitz, *Revolution Politicks*, p. 56.

7 The Revolution

1. HRO, 44M69/F6/9/22, Richard Burd to Thomas Jervoise, 25 October 1688.
2. T. Lever, *Godolphin*, London, 1952, p. 68.
3. Browning, *Thomas Osborne, Earl of Danby*, pp. 387, 390.
4. See for example Horsford, 'Bishop Compton and the Revolution on 1688'.
5. R. A. Beddard, *'A Kingdom Without a King'*, pp. 19–22.

6. Dillon, *The Last Revolution: 1688 and the Creation of the Modern World*, p. 160.
7. Kenyon, *Robert Spencer, Earl of Sunderland*, p. 225.
8. Jones, *Convergent Forces*, p. 143.
9. 'Lord Del-----'s Speech' in *A Collection of Papers relating to the Present Juncture of Affairs in England*, London, 1688, p. 23.
10. M. Beloff, *Public Order and Popular Disturbance 1660–1714*, London, 1963, pp. 40–3.
11. Miller, *James II*, p. 203.
12. Strickland, *The Lives of the Seven Bishops*, pp. 272–3.
13. Harris, 'London Crowds and the Revolution of 1688', pp. 293–4, 300–1, 308.
14. Lord Powis's house was saved because he had opposed the prosecution of the seven bishops – Ibid., p. 297.
15. Beddard, 'Two Letters from the Tower 1688', pp. 347–52.
16. Horowitz, *Revolution Politicks*, p. 57.
17. Jones, *Convergent Forces*, p. 146.
18. *To the King's Most Excellent Majesty The Most Humble and Faithful Advice of Your Majesty's ever Dutiful Subject and Servant, the Bishop of Durham*, London, 1688.
19. M. McClain, 'The Duke of Beaufort's Defence of Bristol During Monmouth's Rebellion, 1685' in *Yale University Library Gazette*, April 2003.
20. Beechey, 'The Defence of the Church of England in James II's Reign', p. 164.
21. 'His Majesty's most Gracious Answer' in *A Collection of Papers relating to the Present Juncture of Affairs in England*, London, 1688, p. 12.
22. NLW, Coedymaen 1, Ms 75, fol. 2.
23. *The Princess Anne of Denmark's Letter to the Queen*, London, 1688.
24. Browning, *Thomas Osborne, Earl of Danby*, p. 401.
25. Evans, 'Yorkshire and the Revolution of 1688', pp. 274–80.
26. 'Modest Vindication of the Petition of the Lords...for the Calling of a Free Parliament' in *A Collection of Papers relating to the Present Juncture of Affairs in England*, London, 1688, p. 14.
27. Kenyon, *Robert Spencer, Earl of Sunderland*, p. 226.
28. NLW, Coedymaen 1, Ms 75, fol. 5.
29. Kenyon, *Robert Spencer, Earl of Sunderland*, p. 228.
30. Lever, *Godolphin*, p. 68.
31. NLW, Kemyes-Tynte Ms c. 124.
32. Sachse, 'The Mob and the Revolution of 1688', pp. 31–2, 35.
33. NLW, Coedymaen 1, Ms 75, fol. 7.
34. Ibid., fol. 9.
35. R. Beddard, 'The Loyalist Opposition in the Interregnum: A Letter of Dr Francis Turner, Bishop of Ely, on the Revolution of 1688' in *Bulletin of the Institute of Historical Research*, vol. 40, 1967, p. 106.
36. Miller, *James II*, p. 207.
37. *London Gazette*, 13 December 1688.
38. Quoted in Beddard, 'The Guildhall Declaration', pp. 406, 414.
39. Ibid., p. 416.
40. Rix, *The Diary and Autobiography of Edmund Bohun*, p. 82. Bohun disliked praying for James thereafter, saying 'we desired him no more'. NLW, Coedymaen 1, Ms 75, fol. 15.
41. NLW, Coedymaen 1, Ms 75, fol. 18.

42. *The Autobiography of Symon Patrick*, p. 139.
43. Quoted in Beddard, *'A Kingdom Without a King'*, p. 145.
44. Beddard, 'The Unexpected Whig Revolution of 1688', pp. 15–16.
45. Storey, *Two East Anglian Diaries*, p. 176.
46. Dillon, *The Last Revolution: 1688 and the Creation of the Modern World*, p. 184.
47. Beddard, 'The Loyalist Opposition in the Interregnum', p. 107.
48. Dillon, *The Last Revolution: 1688 and the Creation of the Modern World*, p. 184.
49. Browning, *Thomas Osborne, Earl of Danby*, p. 423.
50. Beddard, 'The Loyalist Opposition in the Interregnum', p. 104.
51. 'The Letter to a Friend, advising in this Extraordinary Juncture how to Free the Nation from Slavery for ever' in *A Sixth Collection of Papers relating to the Present Juncture of Affairs in England*, London, 1688, p. 13.
52. 'A Word to the Wise for Settling the Government' in *A Sixth Collection of Papers relating to the Present Juncture of Affairs in England*, London, 1688, p. 22.
53. 'A Short Historical Account touching the Succession of the Crown' in *A Sixth Collection of Papers*, London, 1688, pp. 25–8.
54. Hallam, vol. iii, p. 129.
55. Beddard, 'The Loyalist Opposition in the Interregnum', p. 107.
56. NLW, Ms 17015D, Llangibby Castle Collection A60, fols 36–45.
57. Browning, *Thomas Osborne, Earl of Danby*, pp. 426–7.
58. *A Letter from a Bishop to a Lord of his Friends*, London, 1689, pp. 1–2.
59. Browning, *Thomas Osborne, Earl of Danby*, p. 430.
60. Strickland, *The Lives of the Seven Bishops*, p. 352.
61. Ibid., p. 78.
62. Ibid., p. 76 fn. This was their last meeting. William Stanley, Princess Mary's chaplain wrote to George Hickes in May 1713: 'I do not remember that I ever heard that the late good Archbishop Sancroft was thought to have invited the Prince of Orange over into England. If any one did charge him with it, I believe it was without grounds. All that I can say as to the matter is that, Anno. 1687, when I came into England from Holland, I confess I did desire the archbishop to write to the Princess of Orange, on whom I had the honour to attend, to encourage her still to give countenance to the Church of England; but he was pleased not to write to her. And afterwards, when we were come over into England, and a report being spread abroad that some of the lords spiritual, as well as temporal, had invited the Prince of Orange into England, in my communing with the archbishop, I remember he said to me, "I am now glad I did not write to the princess, as you desired, for if I had written to her, they would have said that I had sent to invite them over."'
63. Sancroft had previously incurred Burnet's enmity by refusing to sign an order granting him access to the Cottonian collection of historical MSS. John Evelyn's complaints of the loss he had sustained in consequence of having rashly lent some of the autograph letters of Mary, Queen of Scots, to Dr Burnet, afforded cogent reason to Sancroft for that exclusion.
64. Strickland, *The Lives of the Seven Bishops*, p. 82.
65. Bodleian Library, Tanner MSS, vol. xxvi.
66. D'Oyly, pp. 326–41.
67. 'Several Queries relation to the Present Proceedings In Parliament; more especially recommended to the Consideration of the Bishops' in *An Eighth*

Collection of Papers relating to the Present Juncture of Affairs in England, London, 1689, p. 9.
68. 'A Protestant Precedent offer'd to the Bishops for the Exclusion of K. James the Second' in *An Eighth Collection of Papers*, London, 1689.
69. 'The Bishops Reasons to Queen Elizabeth for taking off the Queen of Scots...' in *An Eighth Collection of Papers*, London, 1689.
70. Beddard, 'The Loyalist Opposition in the Interregnum', p. 108.
71. *The Confinement of the Seven Bishops*, London, 1689.
72. *A Friendly Debate between Dr Kingsman a dissatisfield clergyman and Gratianus Trimmer, a Neighbour Minister*, London, 1689, p. 8.
73. *An Elegy on the Death of that Worthy Prelate the Right Reverend Father in God, Dr John Lake, Late Lord Bishop of Chichester (one of the Seven Bishops who were prisoners in the Tower)...*, London, 1689.
74. Strickland, *The Lives of the Seven Bishops*, pp. 194–5.
75. Hopkins, 'Francis Turner'.
76. Strickland, *The Lives of the Seven Bishops*, pp. 193–4.
77. Ibid., p. 195.
78. W. A. Speck, 'Henry Hyde, Earl of Clarendon', *ODNB*, Oxford, 2004.
79. Strickland, *The Lives of the Seven Bishops*, pp. 217–19.
80. J. S. Sidebotham, *Memorials of King's School, Canterbury*, London, 1865.
81. Strickland, *The Lives of the Seven Bishops*, pp. 274–5.
82. Marshall, 'Thomas Ken'.
83. Strickland, *The Lives of the Seven Bishops*, p. 276.
84. Ibid., p. 277.
85. Marshall, 'Thomas Ken'.
86. Strickland, *The Lives of the Seven Bishops*, p. 288.
87. Not to be confused with Bishop William Lloyd of St Asaph and then Worcester, who was one of the seven.
88. Strickland, *The Lives of the Seven Bishops*, p. 296.
89. Longleat House, where Ken was living with Lord Weymouth.
90. Strickland, *The Lives of the Seven Bishops*, pp. 304–5.
91. Mullett, 'William Lloyd'.

Conclusion

1. Gutch, *Collectanea curiosa*, vol. 1, pp. 370–3. This speech was written in Sancroft's handwriting and therefore likely to represent his views also.
2. Straka, *The Revolution of 1688*, p. 87.
3. Kenyon, *Robert Spencer, Earl of Sunderland*, p. 234.
4. Jones, *Convergent Forces*, p. 130.
5. Ryle, 'James II and the Seven Bishops'.
6. 'Popish Treaties not to be rely'd on: In a Letter from a Gentleman at York, to his friend in the Prince of Orange's Camp...' in *A Third Collection of Papers relating to the Present Juncture of Affairs in England*, London, 1688, p. 32.
7. 'The hard Case of Protestant Subjects under the Dominion of a Popish Prince' in *A Fifth Collection of Papers relating to the Present Juncture of Affairs in England*, London, 1688, pp. 3, 12.

8. 'The Seasonable Queries' in *A Fifth Collection of Papers relating to the Present Juncture of Affairs in England*, London, 1688, p. 33.
9. 'An Account of the irregular Actions of the Papists in the Reign of King James the Second...' in *A Seventh Collection of Papers relating to the Present Juncture of Affairs in England*, London, 1688, pp. 5–6.
10. Ryle, 'James II and the Seven Bishops'.
11. 'The Reasons of the Suddenness of the Change in England' in *The Twelfth and Last Collection of Papers relating to the Present Juncture of Affairs in England*, London, 1689, pp. 7–9.
12. Jones, *Convergent Forces*, p. 141.
13. 'The Reasons of the Suddenness of the Change in England', pp. 10–13.
14. W. Gibson, 'Dissenters, Anglicans and the Glorious Revolution: *The Collection of Cases*'.
15. W. Johnston, 'Revelation and the Revolution of 1688–1689', p. 355.
16. Ibid., p. 364.
17. S. Patrick, *A Sermon Preached in the Chapel of St James's, before his highness the Prince of Orange, the 20th January 1688*, London, 1689, p. 34. The date is old style.
18. *A King or No King*, 1689, Pepys V, 78.
19. *The Lord Chancellors Villanies*, 1689, Pepys II, 288.
20. *A Letter from the Bishop of Rochester to the Right Honourable The Earl of Dorset and Middlesex, Lord Chamberlain of His Majesty's Household concerning his sitting in the Late Ecclesiastical Commission*, London, 1689.
21. *The Bishop of Rochester's Second Letter to the Right Honourable The Earl of Dorset and Middlesex, Lord Chamberlain of His Majesty's Household*, London, 1689, pp. 2–4.
22. Ibid., p. 15.
23. Ibid., p. 17.
24. Havighurst, 'James II and the Twelve Men in Scarlet', p. 545.
25. Harris, 'London Crowds and the Revolution of 1688', pp. 318–26.
26. W. Gibson, 'The Limits of the Confessional State: Electoral Religion in the Reign of Charles II'.
27. Dillon, *The Last Revolution: 1688 and the Creation of the Modern World*, p. 239.
28. Ryle, 'James II and the Seven Bishops'.
29. Spurr, *The Restoration Church of England*, p. 379.

Bibliography

Primary sources

Beinecke Library, Yale University, Osborn Collection, William Bagot letters, FB190.
Bodleian Library, Oxford, Tanner Mss, 28, 29, 32, 57, 62, 66, Rawlinson letters 98–99.
Bristol Record Office, P/StJ/ChW/1b.
British Library, Add. Mss, 34, 510, f. 133.
Derbyshire Record Office, Gell Papers, D258/17/31/58.
Dorset Record Office, Poole Borough Papers, DC/PL/A/1/16.
Hampshire Record Office, 44M69/F6/9/22, Richard Burd to Thomas Jervoise 25 October, 1688.
    9M73/G247/6. Edward Clarke Letter.
Huntington Library, San Marino, California, Ellesmere Ms 8554, the Earl of Huntingdon's copy of proceedings against Compton.
    'Memorandum for Those That Go into the Country to Dispose the Corporations to a Good Election for Members of Parliament. To Be Read by Them Often.' October 1688.
    Hastings Papers, Religious Box 2, folder 9, Draft defence of Huntingdon.
Lambeth Palace Library, VX 1B 2g/2, box 1, deposition of Hugo Powell.
Lancashire Record Office, Kenyon of Peel Mss, DDKE/acc. 7840 Lancashire Trials November 23–5, February 1694/5.
London Metropolitan Archives, Middlesex Sessions of the Peace, MJ/SP/1692, file 3.
National Library of Wales, Coedymaen Ms, 1, ff. 41, 42, 43, 44.
    Kemys-Tynte Ms, c. 124.
    Plas Yn Cefn Ms 2939.
    Brongyntyn Ms 2.
    Ms 17015D, Llangibby Castle Collection.

Printed primary sources

R. A. Beddard, *'A Kingdom without a King' The Journal of the Provisional Government in the Revolution of 1688*, Oxford, 1988.
E. S. de Beer, ed., *The Diary of John Evelyn*, Oxford, 1955.
A. Browning, ed., *Memoirs of Sir John Reresby*, London, 1991 (second edition with notes by M. Geiter and W. Speck).
H. Compton, *The Bishop of London's Seventh Letter of The Conference With His Clergy, held in the year 1686*, London, 1690.
G. de Forest Lord, E. F. Mengel and H. H. Schless, eds, *Poems on Affairs of State*, New Haven, CT, 1963–75.

H. C. Foxcroft *A Supplement to Burnet's History of My Own Time, Derived from His Original Memoirs, His Autobiography, His Letters to Admiral Herbert and His Private Meditations, All Hitherto Unpublished*, Oxford, 1902.

J. Gutch, ed., *Collectanea curiosa; or miscellaneous tracts, relating to the history and antiquities of England and Ireland, and a variety of other Subjects. Chiefly collected and now first published, from the manuscripts of Archbishop Sancroft...*, Oxford, 1781, 2 vols.

H. Hallam, *Constitutional History*, London, 1861–3.

Historical Manuscripts Commission. Report on the Manuscripts of the Duke of Buccleuch and Queensberry..., vol. 2, 1926.

J. Horsfall Turner, ed., *The Works of Oliver Heywood*, London, 1882–5, 4 vols.

T. B. Howell, ed., *Complete Collection of State Trials and Proceedings for High Treason and Other Crimes and Misdemeanours, from the Earliest Period to the Year 1783*, London, 1816, 34 vols.

Henry Howitz, ed., *The Parliamentary Diary of Narcissus Luttrell, 1691–1693*, Oxford, 1972.

J. Hunter, ed., *The Diary of Dr Thomas Cartwright, Bishop of Chester*, Camden Society, 1843.

A. Ivatt, ed., *Memoirs and Travels of Sir John Reresby*, London, 1904.

J. R. Macgrath, *The Flemings in Oxford*, vol. 2, Oxford Historical Society, vol. LXII, 1913.

Robert H. Murray, ed., *The Journal of John Stevens Containing a Brief Account of the War in Ireland 1689–1691*, Oxford, 1912.

S. W. Rix, ed., *The Diary and Autobiography of Edmund Bohun*, Beccles, 1853.

A. Rumble, and D. Dimmer (compilers), edited by C. S. Knighton, *Calendar of State Papers Domestic Series, of the Reign of Anne Preserved in the Public Record Office*, vol. iv, 1705–6, Woodbridge: Boydell Press/The National Archives, 2006.

W. J. Smith, ed., *The Herbert Correspondence: The Sixteenth and Seventeenth Century Letters of the Herberts of Chirbury*, Cardiff, 1968.

M. Storey, ed., *Two East Anglian Diaries, 1641–1729*, Suffolk Records Soc., vol. XXXVI, 1994.

A. Strickland, *The Lives of the Seven Bishops Committed to the Tower in 1688 Enriched and Illustrated with Personal Letters, Now First Published, from the Bodleian Library*, London, 1866.

W. Stubbs, *Registrum Sacrum Anglicanum*, Oxford, 1858.

A. Taylor, ed., *The Works of Simon Patrick*, London, 1858.

E. M. Thompson, ed., *The Letters of Humphrey Prideaux to John Ellis 1674–1722*, Camden Society, 1875.

F. P. Verney, and M. M. Verney, eds, *Memoirs of the Verney Family, 1642–96*, London, 1892–9, 4 vols.

D. H. Woodforde, ed., *Woodforde Diary and Papers*, London, 1932.

Contemporary sources: Anonymous

'An Account of the irregular Actions of the Papists in the Reign of King James the Second...' in *A Seventh Collection of Papers relating to the Present Juncture of Affairs in England*, London, 1688.

An Account of the Late Proposals of the Archbishop of Canterbury with some other Bishops, to his Majesty in a Letter to M. B. Esq, London, 1688.

An Account of the Proceedings at the King's Bench Bar at Westminster Hall Against the Seven Bishops, London, 16 June 1688.

An Address to His Grace the Lord Archbishop of Canterbury, and the Right Reverend the Bishops Upon their Account of their Late Petition, by a true member of the Church of England, London, 1688.

An Answer to a Paper importing a Petition of the Archbishop of Canterbury and Six other Bishops, to His Majesty, touching their not Distributing and Publishing the Late Declaration for Liberty of Conscience, London, 1688.

The Articles Recommended by the Archbishop of Canterbury to all Bishops within his Metropolitan Jurisdiction, the 16th of July, 1688, London, 1688.

Ballad to the Tune of 'Couragio', London, 1688.

The Bishop of Rochester's Second Letter to the Right Honourable The Earl of Dorset and Middlesex, Lord Chamberlain of His Majesty's Household, London, 1689.

'The Bishops Reasons to Queen Elizabeth for taking off the Queen of Scots...' in *An Eighth Collection of Papers relating to the Present Juncture of Affairs in England*, London, 1689.

Bishop Sancroft's Ghost with his Prophesie of the Times, And the Approach of the Anti-Christ, London, 1712.

The Clergy's Late Carriage to the King Considered in a Letter to a Friend, London, 1688.

A Collection of Cases, and other Discourses, lately written to recover Dissenters to the Communion of the Church of England, by some divines of the City of London, London, 1685, 2 vols.

A Collection of the Most Remarkable and Interesting Trials. Particularly of those Persons who have forfeited their lives to the injured laws of their country. In which the most remarkable of the State Trials will be included, London, 1775–6, 2 vols.

The Confinement of the Seven Bishops, London, 1688.

The Convocation, London, 1688.

A Copy of An Address to the King by the Bishop of Oxon to be subscribed by the Clergy of His Diocese; with the Reasons for the Subscription to the Address and the reasons against it, London, 1687.

The Countrey-Minister's Reflections on the City Minister's Letter to his Friend, Shewing the Reasons why We cannot Read the King's Declaration in Our Churches, London, 1688.

'Lord Del-----'s Speech' in *A Collection of Papers Relating to the Present Juncture of Affairs in England*, London, 1688.

The Dialogue, London, 1688.

A Dialogue between the ArchB of C. and the Bishop of Heref. Containing the True reasons why the Bishops could not read the Declaration, London, 1688.

An Elegy on the Death of that Worthy Prelate the Right Reverend Father in God, Dr John Lake, Late Lord Bishop of Chichester (one of the Seven Bishops who were prisoners in the Tower)..., London, 1689.

An Exact Account of the Whole Proceedings Against the Right Reverend Father in God, Henry Lord Bishop of London before the Lord Chancellor and the Other Ecclesiastical Commissioners, London, 1688.

The Examination of the Bishops Upon their Refusal of Reading His Majesty's Most Gracious Declaration; And the Nonconcurrence of the Church of England, In Repeal of the Penal Laws and Test Fully Debated and Argued, London, 1688.

A Friendly Debate between Dr Kingsman a dissatisfield clergyman and Gratianus Trimmer, a Neighbour Minister, London, 1689.
Great and Good News to the Church of England, London, 1705.
Great and Good News for the Church of England If They Please to Accept thereof: Or the Latitudinarian Christians Most Humble Address and Advice, London, 1688.
'The hard Case of Protestant Subjects under the Dominion of a Popish Prince' in *A Fifth Collection of Papers relating to the Present Juncture of Affairs in England*, London, 1688.
His Majesty's Declaration to all His Loving Subjects, Touching the Causes and Reasons That Moved him to Dissolve the Two Last Parliaments, London, 1681.
'His Majesty's most Gracious Answer' in *A Collection of Papers relating to the Present Juncture of Affairs in England*, London, 1688.
The History of King James's Ecclesiastical Commission: Containing all the Proceedings against The Lord Bishop of London; Dr Sharp, now Archbishop of York; Magdalen-College in Oxford; The University of Cambridge; The Charter-House at London and The Seven Bishops, London, 1711.
A King or No King, London, 1689.
To the King's Most Excellent Majesty: The Most Humble and Faithful Advice of Your Majesty's ever Dutiful Subject and Servant, the Bishop of Durham, London, 1688.
The Learned and Loyal Abraham Cowley's Definition of a Tyrant...in his Discourse concerning the Government of Oliver Cromwell, London, 1688.
A Letter from a Bishop to a Lord of his Friends, London, 1689.
A Letter from the Bishop of Rochester to the Right Honourable The Earl of Dorset and Middlesex, Lord Chamberlain of His Majesty's Household concerning his sitting in the Late Ecclesiastical Commission, London, 1689.
A Letter from a Clergyman in the City to his friend in the country, Containing his Reasons for not Reading the Declaration, London, 1688.
Letter from a Country Curate to Mr Henry Care in Defence of the Seven Bishops..., London, 1688.
Letter to a Dissenter, London, 1687.
'The Letter to a Friend, advising in this Extraordinary Juncture how to Free the Nation from Slavery for ever' in *A Sixth Collection of Papers relating to the Present Juncture of Affairs in England*, London, 1688.
A Letter of several French Ministers Fled into Germany upon the account of the Persecution in France, to such of their Brethren in England as Approved the King's Declaration touching Liberty of Conscience, translated from the Original in French, London, 1688.
'Letters of Bishop Levinze to Thomas Choldmondley', *Proceedings of the Manx Society*, vols IV, VII and IX.
Lex ignea, or, The School of righteousness a sermon preach'd before the King, Octob. 10. 1666 at the solemn fast appointed for the late fire in London by William Sancroft, London, 1666.
The Lord Chancellors Villanies, London, 1689.
'A Memorial of the Protestants of the Church of England to their Royal Highnesses the Prince and Princess of Orange' in *A Collection of Papers relating to the Present Juncture of Affairs in England*, London, 1688.
The Minister's Reasons for His Not reading the King's Declaration, Friendly Debated by a Dissenter, London, 1688.

'Modest Vindication of the Petition of the Lords...for the Calling of a Free Parliament' in *A Collection of Papers relating to the Present Juncture of Affairs in England*, London, 1688.
Nahash Revived: Or the Church of England's Love to Dissenters and Loyalty to their Prince..., London, 1688.
'On the Seven Bishops' in *A Collection of Epigrams*, London, 1727.
A Pastoral Letter from the Four Catholic Bishops to the Lay Catholics of England, London, 1688.
A Poem on the Deponents Concerning the Birth of the Prince of Wales, London, 1688.
'Popish Treaties not to be rely'd on: In a Letter from a Gentleman at York, to his friend in the Prince of Orange's Camp...' in *A Third Collection of Papers relating to the Present Juncture of Affairs in England*, London, 1688.
The Prince of Orange Welcome to London, London, 1688.
'The Prince of Orange's first Declaration from the Hague...with his Highnesses Additional Declaration from the Hague, October 24 1688' in *A Fourth Collection of Papers relating to the Present Juncture of Affairs in England*, London, 1688.
The Princess Anne of Denmark's Letter to the Queen, London, 1688.
The Proceedings and Tryal in the Case of the Most Reverend Father in God, William, Lord Archbishop of Canterbury, ...in the Court of the King's Bench, London, 1739.
The Proceedings and Tryal in the Case of the most Reverend Father in God, William, Lord Archbishop of Canterbury and the Right Reverend Fathers in God, William, Lord Bishop of St. Asaph, Francis, Lord Bishop of Ely, John, Lord Bishop of Winchester, Thomas, Lord Bishop of Bath and Wells, Thomas, Lord Bishop of Peterborough, and Jonathan, Lord Bishop of Bristol, in the Court of Kings-Bench at Westminster in Trinity-term in the fourth year of the reign of King James the Second, Annoque Dom. 1688, London, 1689.
A Proclamation by the King for Suppressing and Preventing Seditious and Unlicenced Books and Pamphlets [broadsheet], 10 February 1687/8.
A Prophylactick from Disloyalty in these Perilous Times. In a Letter to the Right Honourable and Right Reverend Father in God Herbert, By Divine Providence, Lord Bishop of Hereford, London, 1688.
'A Protestant Precedent offer'd to the Bishops for the Exclusion of K. James the Second' in *An Eighth Collection of Papers relating to the Present Juncture of Affairs in England*, London, 1689.
'The Reasons of the Suddenness of the Change in England' in *The Twelfth and Last Collection of Papers relating to the Present Juncture of Affairs in England*, London, 1689.
'To the Right Honourable My Lords of his Majesty's Commission Ecclesiastical' in *A Fourth Collection of Papers relating to the Present Juncture of Affairs in England*, London, 1688.
'The Seasonable Queries' in *A Fifth Collection of Papers relating to the Present Juncture of Affairs in England*, London, 1688.
The Sentence of Samuel Johnson, at the Kings-Bench-Barr at Westminster..., London, 1686.
'Several Queries relation to the Present Proceedings in Parliament; more especially recommended to the Consideration of the Bishops' in *An Eighth Collection of Papers relating to the Present Juncture of Affairs in England*, London, 1689.
The Sham Prince Expos'd In A Dialogue between the Pope's Nuncio and the Bricklayer's Wife, Nurse to the Supposed Prince of Wales, London, 1688.

'A Short Historical Account touching the Succession of the Crown' in *A Sixth Collection of Papers relating to the Present Juncture of Affairs in England*, London, 1688.

'Some Account of the Humble Application of the Pious and Noble Prelate, Henry Lord Bishop of London, with the Reverend Clergy of the City and some of the Dissenting Ministers in it, to the Illustrious Prince William Henry...' in *A Sixth Collection of Papers relating to the Present Juncture of Affairs in England*, London, 1688.

Some Queries Humbly Offered to the Lord Archbishop of Canterbury and the Six Other Bishops concerning the English Reformation And the Thirty Nine Articles of the Church of England, London, 1688.

Supposed Prince of Wales, London, 1688.

Toleration Tolerated, or A Late Learned Bishop's Opinion Concerning Toleration of Religion, London, 1688.

Trayal of the Seven Bishops, With a Preface Shewing the Present Danger of our Religion and Liberties..., London, 1713.

Two Plain Words to the Clergy: Or an Admonition to Peace and Concord at this Juncture, London, 1688.

A View of the Popish Plot, London, 1689.

What has been may be: Or A View of a Popish and an Arbitrary Government... To which is added the Twenty-One Conclusions Further Demonstrating the Schism of the Church of England, London, 1688.

'A Word to the Wise for Settling the Government' in *A Sixth Collection of Papers relating to the Present Juncture of Affairs in England*, London, 1688.

Contemporary sources

E. Bohun, *The Doctrine of Non-resistance or Passive Obedience no way concerned in the Controversies now depending between Williamites and the Jacobites*, London, 1689.

D. Defoe, *A New Test of the Church of England's Loyalty: Or Whiggish Loyalty and Church Loyalty Compar'd*, London, 1702.

J. Ellerby, *Doctrine of Passive Obedience*, London, 1685/6.

E. Ellis, *A Clergyman of the Church of England: His Vindication of Himself for Reading His Majesties Late Declaration*, London, 1688.

D. Jones, *The life of James II. late King of England. Containing an account of His birth, education, religion, and enterprizes,... till his dethronement. With the various struggles made since for his restoration;... and the particulars of his death. The whole intermix'd with divers original papers*, London, 1703.

T. Ken, *All Glory be to God: Thomas, Unworthy Bishop of Bath and Wells...*, London, 1688.

—— *A Pastoral Letter from the Bishop of Bath and Wells to his Clergy Concerning their Behaviour During Lent*, London, 1688.

A. Marvel, *Mr Smirke, or the Divine in Mode...*, London, 1676.

S. Patrick, *A Sermon Preached in the Chapel of St James's, before his highness The Prince of Orange, the 20th January 1688*, London, 1689.

W. Sancroft, *A sermon preached in S. Peter's Westminster, on the first sunday in Advent, at the consecration of the Right Reverend Fathers in God, John Lord Bishop*

of Durham, William Lord Bishop of S. David's, Beniamin L. Bishop of Peterborough, Hugh Lord Bishop of Landaff, Richard Lord Bishop of Carlisle, Brian Lord Bishop of Chester, and John Lord Bishop of Exceter, London, 1660.
T. Tenison, *An Argument for Union...*, London, 1683.
W. Thomas, *The Mammon of Unrighteousness Detected and Purified in a Sermon Preached in the Cathedral Church of Worcester on Sunday the Nineteenth of August, 1688*, London, 1688.
F. Turner, *A Sermon Preached Before their Majesties K. James II and Q. Mary at their Coronation in Westminster Abby, April 23. 1685*, London, 1685.

Secondary sources

M. Ashley, *The Glorious Revolution of 1688*, London, 1966.
M. Barone, *Our First Revolution, the Remarkable British Upheaval That Inspired America's Founding Fathers*, New York, 2007.
A. Barrow, *The Flesh Is Weak*, London, 1980.
M. Beloff, *Public Order and Popular Disturbance 1660–1714*, London, 1963.
G. V. Bennett, *White Kennett, 1660–1728*, London, 1957.
J. Black, *A System of Ambition? British Foreign Policy 1660–1793*, London, 1991.
A. Browning, *Thomas Osborne, Earl of Danby and Duke of Leeds, 1632–1712*, Glasgow, 1951, 2 vols.
E. Carpenter, *The Protestant Bishop*, London, 1956.
—— *Thomas Tenison*, London, 1948.
J. Carswell, *The Descent on England*, London, 1969.
J. C. D. Clark, *English Society 1660–1832*, Second edition, Cambridge, 2000.
J. S. Clarke, *The Life of James II by Himself*, London, 1816, 2 vols.
T. E. S. Clarke and H. C. Foxcroft, *A Life of Gilbert Burnet... with an Introduction by C. H. Firth*, Cambridge, 1907.
T. Claydon, *Europe and the Making of England 1660–1760*, Cambridge, 2007.
—— *William III and the Godly Revolution*, Cambridge, 1996.
A. Coleby, *Central Government and the Localities: Hampshire 1649–1689*, Cambridge, 1987.
B. Coward, *The Stuart Age: A History of England 1603–1714*, London, 1980.
G. S. de Krey, *Restoration and Revolution in Britain: A Political History of the Era of Charles II and the Glorious Revolution*, London, 2007.
P. Dillon, *The Last Revolution: 1688 and the Creation of the Modern World*, London, 2006.
G. D'Oyly, *The Life of William Sancroft*, London, 1840.
—— *The Life of William Sancroft, archbishop of Canterbury*, London, 1821, 2 vols.
G. Every, *The High Church Tradition*, London, 1956.
W. Gibson, *Religion and the Enlightenment 1600–1800, Conflict and the Rise of Civic Humanism in Taunton*, Oxford, 2006.
M. Goldie, *Roger Morrice and the Puritan Whigs, the Entering Book of Roger Morrice, 1677–1691*, Woodbridge, 2007.
E. Green, *The March of William of Orange through Somerset with a Notice of Other Local Events in the time of King James II*, London, 1892.
T. Harris, *Revolution: The Great Crisis of the British Monarchy, 1685–1720*, London, 2006.

A. T. Hart, *The Life and Times of John Sharp, Archbishop of York*, London, 1949.
—— *William Lloyd 1627–1717*, London, 1952.
B. W. Hill, *The Growth of Parliamentary Parties 1689–1742*, London, 1976.
C. Hill, *The Century of Revolution 1603–1714*, London, 1961.
H. Horowitz, *Revolution Politicks, the Career of Daniel Finch, Second Earl of Nottingham 1647–1730*, Cambridge, 1968.
J. Israel, ed., *Anglo-Dutch Moment, Essays on the Glorious Revolution and Its World Impact*, Cambridge, 2003.
G. H. Jones, *Convergent Forces: Immediate Causes of the Revolution of 1688*, Ames, IA, 1990.
J. R. Jones, *The Revolution of 1688 in England*, London, 1972.
G. W. Keeton, *Lord Chancellor Jeffreys and the Stuart Cause*, London, 1965.
J. P. Kenyon, *The Nobility in the Revolution of 1688*, University of Hull, 1963.
—— *Revolution Principles*, Cambridge, 1990.
—— *Robert Spencer, Earl of Sunderland 1641–1702*, London, 1958.
D. Lacey, *Dissent and Parliamentary Politics in England 1661–1689: A Study in the Perpetuation and Tempering of Parliamentarianism*, New Brunswick, NJ, 1969.
T. Lever, *Godolphin*, London, 1952.
H. Lucock, *The Bishops in the Tower*, London, 1896.
T. B. Macaulay, *History of England*, London, 1848.
J. Miller, *James II a Study in Kingship*, Hove, 1977.
—— *Popery and Politics in England 1660–1688*, Cambridge, 1973.
M. Mullett, *James II and English Politics 1678–1688*, London, 1994.
J. H. Overton, *The Nonjurors: Their Lives, Principles, and Writings*, London, 1902.
S. Pincus, *England's Glorious Revolution*, London, 2006.
L. Pinkham, *William III and the Respectable Revolution*, Cambridge, MA, 1954.
H. A. L. Rice, *Thomas Ken, Bishop and Nonjuror*, London, 1958.
W. L. Sachse, *Lord Somers, a Political Portrait*, Manchester, 1975.
L. Schwoerer, *The Declaration of Rights, 1689*, Baltimore, MD, 1981.
—— *The Revolution of 1688–9: Changing Perspectives*, Cambridge, 1991.
J. S. Sidebotham, *Memorials of King's School, Canterbury*, London, 1865.
M. G. Smith, *Fighting Joshua, Sir Jonathan Trelawny 1650–1721*, Redruth, 1985.
W. A. Speck, *Profiles in Power: James II*, London, 2002.
—— *Reluctant Revolutionaries: Englishmen and the Revolution of 1688*, London, 1988.
W. M. Spellman, *The Latitudinarians and the Church of England 1660–1700*, Athens, GA, 1993.
J. Spurr, *The Restoration Church of England 1646–1689*, London and New Haven, CT, 1991.
G. M. Straka, *Anglican Reaction to the Revolution of 1688*, Wisconsin, 1962.
N. Sykes, *William Wake*, Cambridge, 1957, 2 vols.
G. M. Trevelyan, *The English Revolution 1688–1689*, Oxford, 1950.
E. Vallance, *The Glorious Revolution: 1688 – Britain's Fight for Liberty*, London, 2007.
J. R. Western, *Monarchy and Revolution: The English State in the 1680s*, London, 1972.
C. E. Whitting, *Nathaniel, Lord Crewe, Bishop of Durham 1674–1721*, London, 1940.

Articles and essays

J. Barry, 'Exeter in 1688: The Trial of the Seven Bishops' in *Devon and Cornwall Notes and Queries*, vol. 38, 1996.

R. A. Beddard, 'The Loyalist Opposition in the Interregnum: A Letter of Dr Francis Turner, Bishop of Ely, on the Revolution of 1688' in *Bulletin of the Institute of Historical Research*, vol. 40, 1967.

—— 'Observations of a London Clergyman on the Revolution of 1688–9: Being an Except from the Autobiography of Dr William Wake' in *Guildhall Miscellany*, vol. 2, 1967.

—— 'Two Letters from the Tower 1688' in *Notes and Queries* ccxxix, September 1984.

—— 'The Unexpected Whig Revolution of 1688' in Beddard, ed., *The Revolutions of 1688*, Oxford, 1991.

—— 'William Sancroft' in *Oxford Dictionary of National Biography*, Oxford, 2005.

G. V. Bennett, 'Loyalist Oxford and the Revolution' in L. S. Sutherland and L. G. Mitchell, eds, *The History of the University of Oxford, vol 5: The Eighteenth Century*, Oxford, 1986.

—— 'The Seven Bishops: A Reconsideration' in D. Baker, ed., *Studies in Church History*, vol. 15, 1978.

R. Bretton, 'Bishop John Lake', *Transactions of the Halifax Antiquarian Society*, Halifax, 1968.

A. M. Coleby, 'Jonathan Trelawny', *Oxford Dictionary of National Biography*, Oxford, 2004.

E. Cruickshanks, 'The Revolution and the Localities: Examples of Loyalty to James II' in E. Cruickshanks, ed., *By Force Or by Default? The Revolution of 1688–1689*, Edinburgh, 1989.

D. Davies, 'James II, William of Orange and the Admirals' in E. Cruickshanks, ed., *By Force Or by Default? The Revolution of 1688–1689*, Edinburgh, 1989.

G. S. de Krey, '" Reformation and Arbitrary Government": London Dissenters and James II's Polity of Toleration, 1687–1688' in J. McElligott, ed., *Fear, Exclusion and Revolution, Roger Morrice and Britain in the 1680s*, Aldershot, 2006.

H. T. Dickinson, 'The Eighteenth Century Debate on the "Glorious Revolution"', *History*, vol. 28, 1976.

D. Dixon, '*Godden v Hales* Revisited – James II and the Dispensing Power' in *Journal of Legal History*, vol. 27, no. 2, 2006.

A. M. Evans, 'Yorkshire and the Revolution of 1688' in *Yorkshire Archaeological Journal*, vol. 29, 1929.

R. H. George, 'The Charters Granted to English Parliamentary Corporations in 1688' in *English Historical Review*, vol. LV, 1940.

W. Gibson, 'Dissenters, Anglicans and the Glorious Revolution: *The Collection of Cases*' in *The Seventeenth Century*, vol. 22, no. 1, 2007.

—— 'The Limits of the Confessional State: Electoral Religion in the Reign of Charles II' in *Historical Journal*, 2008.

—— '"Look toward the Court!" Two Jesus Men in 1688' in *Jesus College Record*, 2008.

M. Goldie, 'The Political Thought of the Anglican Revolution' in Beddard, ed., *The Revolutions of 1688*, Oxford, 1991.

M. Goldie and J. Spurr, 'Politics and the Restoration Parish: Edward Fowler and the Struggle for St Giles Cripplegate' in *English Historical Review*, 1994.

T. Harris, 'London Crowds and the Revolution of 1688' in E. Cruickshanks, ed., *By Force Or by Default? The Revolution of 1688–1689*, Edinburgh, 1989.

A. F. Havighurst, 'James II and the Twelve Men in Scarlet' in *Law Quarterly Review*, vol. 69, 1953.

P. Hopkins, 'Francis Turner', *Oxford Dictionary of National Biography*, Oxford, 2004.

H. Horowitz, '1689 (and all that)' in *Parliamentary History*, vol. 6, 1978.

—— 'Parliament and the Glorious Revolution' in *Bulletin of the Institute of Historical Research*, vol. 47, 1974.

D. H. Horsford, 'Bishop Compton and the Revolution on 1688' in *Journal of Ecclesiastical History*, vol. xxiii, 1972.

W. Johnston, 'Revelation and the Revolution of 1688–1689' in *Historical Journal*, vol. 48, no. 2, 2005.

D. L. Jones, 'The Glorious Revolution in Wales' in *The National Library of Wales Journal*, vol. 26, no. 1, 1989.

J. P. Kenyon, 'The Earl of Sunderland and the Revolution of 1688' in *Cambridge Historical Journal*, vol. 10, 1955.

—— 'Introduction' in E. Cruickshanks, ed., *By Force Or by Default? The Revolution of 1688–1689*, Edinburgh, 1989.

N. Key, 'Comprehension and the Breakdown of Consensus in Restoration Herefordshire' in T. Harris, P. Seaward and M. Goldie, eds, *The Politics of Religion in Restoration England*, Oxford, 1990.

M. McCain, 'The Duke of Beaufort's Defence of Bristol During the Monmouth's Rebellion 1685' in *Yale University Gazette*, April 2003.

A. MacIntyre, 'The College, James II and the Revolution 1687–1688' in *Magdalen College and the Crown*, Oxford, 1988.

W. Marshall, 'Thomas Ken', *Oxford Dictionary of National Biography*, Oxford, 2004.

M. Mendle, 'The "Prints" of the Trials: The Nexus of Politics, Religion, Law and Information in Late Seventeenth Century England' in J. McElligott, ed., *Fear, Exclusion and Revolution, Roger Morrice and Britain in the 1680s*, Aldershot, 2006.

J. Miller, 'James II and Toleration' in E. Cruickshanks, ed., *By Force Or by Default? The Revolution of 1688–1689*, Edinburgh, 1989.

M. Mullett, 'The Politics of Liverpool 1660–1688' in *Transactions of the Historical Society of Lancashire and Cheshire*, vol. 124, 1972.

—— 'Recusants, Dissenters and North-West Politics between the Restoration and the Glorious Revolution' in J. Appleby and P. Dalton, eds, *Government, Religion and Society in Northern England 1000–1700*, Stroud, 1997.

—— 'Thomas White', *Oxford Dictionary of National Biography*, Oxford, 2004.

J. G. A. Pocock, 'The Significance of 1689: Some Reflections on Whig History' in Beddard, ed., *The Revolutions of 1688*, Oxford, 1991.

H. H. Poole, 'John Lake', *Oxford Dictionary of National Biography*, Oxford University Press, 2004.

J. C. Ryle, 'James II and the Seven Bishops' in *Light from Old Times*, London, 1902.

W. L. Sachse, 'The Mob and the Revolution of 1688' in *The Journal of British Studies*, vol. 4, 1964.

Y. Sherwood, 'The God of Abraham and Exceptional States, Or the Early Modern Rise of the Whig/Liberal Bible' in *The Journal of the American Academy of Religion*, vol. 76, no. 2, 2008.
W. A. Speck, 'Henry Hyde, Earl of Clarendon', *Oxford Dictionary of National Biography*, Oxford, 2004.
J. Stoye, 'Europe and the Revolution of 1688' in Beddard, ed., *The Revolutions of 1688*, Oxford, 1991.
S. Taylor, '"Plus Ca Change...?" New Perspectives on the Revolution of 1688' in *Historical Journal*, vol. 37, no. 2, 1994.
R. Thomas, 'The Seven Bishops and Their Petition, 18 May 1688' in *The Journal of Ecclesiastical History*, vol. 12, 1961.
K. Wilson, 'Inventing Revolution: 1688 and Eighteenth Century Popular Politics' in *Journal of British Studies*, vol. 28, 1989.

Theses

C. R. Beechey, 'The Defence of the Church of England in James II's Reign', Lancaster University, MA thesis, 1985.
P. D. Heitzman, 'The Revolution of 1688 in Lancashire and Cheshire', Manchester University, MPhil thesis, 1984.
Roger D. Jones, 'The Invitation to William of Orange in 1688', University of Wales, Cardiff, MA thesis, 1969.
Newton E. Key, 'Politics beyond Parliament: Unity and Party in the Herefordshire Region during the Restoration Period', Cornell University, PhD thesis, 1989.
C. A. Lowe, 'Politics and Religion in Warwickshire during the Reign of James II, 1685–1688', Warwick University, MA thesis, 1992.

Index

Abingdon, Earl of, 70, 124, 166, 170
acquittal of the bishops, 9, 18, 136–9, 141
Addison, Lancelot, 136
Ailesbury, Lord, 169, 171, 173, 176
Alfred, King, ix
Allybone, Judge, 116, 131, 148
Alsopp, Vincent, 66, 81
Anne, Hyde Duchess of York (first wife of James II), 27, 71
Anne, Queen (Princess Anne), 15, 30, 40, 42, 43, 51, 61, 71, 74, 142, 167, 168, 180, 189, 192
anointing, 38
anti-Catholicism, 33, 48, 53, 56–8, 59, 72, 73, 105, 133–9, 147, 164, 168, 171, 196, 197
Archer, Isaac, 103–4, 136, 175
army, 10, 11, 45, 51, 52, 53, 54, 70–1, 133, 137, 140, 142, 162, 165, 166, 167, 170, 172, 173, 174
articles of advice, 155, 156, 195
Arundel, Lord, 55, 71, 78
Ashley, Maurice, 85
assizes, 69
Astray, Sir Samuel, 118

Baber, Sir John, 81, 152
Bacon, Francis, 92
bail, 108, 109, 117
Baldock, Serjeant, 114, 130
Barclay, Robert, 111
Barillon, Paul, d'Amoncourt, 10, 51, 143, 156, 157, 166, 169
Barnstaple, 77
Basing House, siege of, 23
Bath, 41, 51, 86
Bath, Earl of, 69, 70, 166
Bath and Wells, 18, 31, 42, 86, 154, 189, 192
Baxter, Richard, 50, 64, 196
Beaufort, Duke of, 14, 39, 166

Beddard, Robert, 12, 13, 14, 37, 80, 172, 173
Bedford, 135
Bedford, Earl of, 123
Belasyse, Lord, 55, 71, 74, 78
Bennett, G. V., 8, 9, 64
Bentinck, Hans Willem (Lord Portland), 112
Berkeley, Lord, 109
Bertie, Charles, 171
Beveridge, Bishop William, 181, 191
Bill of Rights, 3, 201
Birmingham, 171
Black, Jeremy, 8, 16
Blackheath, 71
Bohun, Edmund, 37, 45, 136, 152, 173
Bonrepaus, François d'Usson de, 67, 152
Bradford, Lord, 70
Bramston, Sir John, 164
Bridgeman, William, 91, 126
Bridgewater, Earl of, 70
Bridport, 77
Bristol, 14, 18, 29, 30, 32, 33, 39, 135, 136, 142, 147, 154, 164, 165, 166, 171
Bristol, Earl of, 162
Brownrigg, Bishop Ralph, 24
Buckinghamshire, 69, 144
Bullingbrook, Lord, 123
Burd, Richard, 162
Burlington, Earl of, 159, 161
Burnet, Bishop Gilbert, 6, 47, 52, 59, 75, 78, 121, 151, 170, 175, 176, 178, 181, 183, 189, 190
Bury St Edmund, 165
Butler, Sir Nicholas, 74

Cambridge, 21, 22, 23, 25, 37, 62, 135, 147, 148, 165, 171, 181
  Emmanuel College, 22, 23, 25
  St John's College, 27
  Sidney Sussex College, 62

244 *Index*

Canterbury, 18, 60, 119, 183
Care, Henry, 66
Carlisle, 29, 65
Carlisle, Earl of, 162
Carnarvon, Earl of, 123
Carswell, John, 14, 91, 141
Carteret, Lord, 124
Cartwright, Bishop Thomas, 24, 59,
    60, 62, 64, 77, 78, 81, 84, 104,
    144, 147, 157
Catherine, Queen, wife of Charles II,
    31, 174
Cave, William, 79
Chandoys, Lord, 124
Chapel royal, 42, 43, 54, 55, 57, 97,
    133, 175
Charles I, King, ix, 6, 20, 21, 23, 34,
    44, 45, 92
Charles II, King, 5, 7, 25, 26, 29, 31,
    32, 36, 38, 39, 47, 48, 49, 51, 62,
    152, 157, 163, 174, 176, 198, 202
Charlett, Arthur, 136
Charterhouse, 37, 60
Cheshire, 164, 166
Chester, 51, 59, 65, 66, 67, 77, 84, 103,
    104, 144, 171
Chesterfield, 173
Chesterfield, Earl of, 123
Chichester, 18, 32, 33, 39, 40, 142,
    143, 154
Cholmodeley, Thomas, 119
Churchill, John (Lord Churchill,
    Duke of Marlborough), 12, 51,
    52, 61, 70, 142, 167, 169, 180, 193,
    198
Civil War, 20, 22, 24, 50
Claggett, William, 53, 79, 80, 150
clandestine networks, 139
Clarendon, Earl of, 12, 14, 18, 36,
    60, 61, 71, 83, 84, 92, 104, 108,
    111, 114, 117, 123, 132, 153, 159,
    166, 167, 169, 170, 173, 175, 180,
    181, 183, 185, 186, 187, 189, 198
Clarendon code, 50, 73, 93
Clark, Edward, 97, 133
Clark, Jonathan, 6
Clerkenwell, 71, 147
Cleveland, 26

'Closeting' campaign, 60–1, 81
Colchester, Lord, 166
Collection of Cases, 79, 83
Commonwealth, 23, 24, 25, 26, 27
Compton, Bishop Henry, 2, 4, 18,
    29, 30, 32, 36, 40, 46, 48, 50, 52,
    53–9, 73, 74, 80, 84, 85, 90, 91,
    111, 112, 118, 120, 141, 142, 151,
    153, 154, 157, 159, 160, 161, 163,
    167, 168, 175, 177, 178, 183, 185,
    186, 198, 199, 200
Compton, Sir Francis, 58
Conventicle Act, 56
Cornbury, Lord, 166, 198
Cornwall, 2, 3, 20, 28, 33, 43, 69, 88,
    137
Cosin, Bishop John, 24
Coventry, Earl of, 59
Cowley, Abraham, 87
Cranmer, Archbishop Thomas, 22, 96
Craven, Lord, 172
Crewe, Bishop Nathaniel, 2, 29, 54,
    57, 58, 66, 68, 81, 103, 144, 157,
    166
Croft, Bishop Herbert, 103, 105, 149
Cromwell, Oliver, 36, 149, 154

D'Adda, Ferdinando, 156
Danby, Earl of, 18, 48, 51, 52, 54, 56,
    60, 61, 76, 78, 115, 120, 123, 140,
    141, 142, 159, 163, 164, 168, 171,
    176, 177, 179, 180, 198
Dartmouth, Lord, 119, 153, 172
David, King, 38–9, 92
Declaration of Indulgence, x, 6, 15,
    19, 37, 64–8, 69, 72, 77, 78–83,
    85–96, 97–109, 113, 115, 116, 118,
    127, 129, 130, 137, 139, 143, 144,
    145, 146, 152, 153, 158, 179, 194,
    198, 201
Defoe, Daniel, 13, 111
Delamere, Lord, 164, 166, 174
Derby, Earl of, 29, 70, 156
Derbyshire, 151
Devon, 104, 163, 168
Devonshire, Earl of, 75, 141, 163
dispensing power, 57, 62, 64, 82, 86,
    88, 105, 127, 128, 129, 139, 155

Dissent, Protestant, 4, 5, 6, 8, 9, 11, 15, 16, 19, 36, 38, 40, 41, 50, 53, 56, 61, 64, 65, 66, 70, 72, 73, 75, 77, 79, 80, 81, 82–96, 98, 105–9, 115, 117, 118, 120, 136, 137, 139, 140, 143, 145, 146, 149, 150, 151, 152, 153, 154, 156, 157, 172, 193, 195, 196, 197, 198, 199, 202
Dorset, 33
Dorset, Earl of, 76, 77, 123
Dover, 54
Dover, Lord, 55, 70, 74, 171
Dryden, John, 38, 89–90
Dublin, 81
Dunblane, Lord, 159
Dunmore, Lord, 156
Durham, 25, 103, 144

Ecclesiastical Commission, 2, 7, 8, 15, 37, 46, 53–9, 62, 63, 67–8, 106, 144, 146, 154, 155, 156, 157, 200, 201
economy, 11
Edinburgh, 27
Edward II, King, 49
Edward the confessor, King, 177
Elizabeth I, Queen, ix, 47, 96, 165
Ellis, Edmund, 104–5
Ellis, Bishop Phillip, 81
Ely, 18, 31, 84
Essex, 77
Europe, 8, 196
Evelyn, John, 42, 56, 57, 64, 98, 110
exclusionism, of James II, 8, 12, 32, 35, 46, 47, 49, 50, 52, 63, 97, 139, 172
Exeter, 32, 56, 135, 136, 137, 142, 163, 166, 167, 193

Fagel, Grand Pensionary, 75
Farmer, Anthony, 63
Farnham Castel, 63
Fauconberge, Viscount, 124
Faversham, Lord, 165, 172, 174
Fell, John, 24
Fenwick, Sir John, 187, 189
Finch, Heneage, 55, 114
Forde Abbey, 163

Fowler, Edward, 53, 79, 83, 84
Foxe, John, 16
Frampton, Bishop Robert, 85, 148, 181, 189
France, 7, 24, 52, 102, 162, 169, 171, 175, 183, 187
Francis, Alban, 62, 196

Gadbury, John, 82, 153
Gainsborough, Earl of, 70
Gell, Sir John, 151
George, Prince of Denmark, 30, 167
Giffard, Bishop Bonaventure, 76, 81
Gisborne, John, 151
Glorious Revolution, ix, 4, 5, 6, 7, 8, 10, 12, 13, 16, 17, 18, 19, 38, 96, 139, 181, 184, 188, 199, 201, 202, 203
Gloucester, 68, 73, 135, 148, 165, 166, 171, 178
Godden, Arthur, 54–5, 86, 116
Godfrey, Sir Edmund Berry, 32
Godolphin, Lord, 156, 163, 169, 192, 193
Goldie, Mark, 14, 15, 16, 17, 136
Goodricke, Sir Henry, 159
Grafton, Duke of, 167, 174
Gregory, Sir William, 55
Grey of Ruthyn, Lord, 124
Grove, Robert, 84
Guildhall meeting, 171–5
Gunning, Bishop Peter, 31
Gunpowder plot, 3, 73, 183
Gwyn, Nell, 31, 72

habeas corpus, 115
Hales, Sir Edward, 54–5, 64, 86, 99, 109, 112, 116, 139, 166
Halifax, Marquis of, 10, 12, 14, 17, 18, 47, 48, 52, 66, 76, 78, 83, 123, 133, 141, 159, 161, 169, 170, 176, 177, 179, 198
Halifax, Yorkshire, 21, 23, 166
Hall, Bishop Timothy, 144, 145
Hampshire, 148, 162
Hanbury, Father Robert, 158
Hanmer, Sir John, 168
Harris, Tim, 16, 17, 156

Hatton, Charles, 167
Hatton, Viscount, 58
Havighurst, Alfred, 12, 55, 139
Hawkins, Francis, 144
Hawles, Sir John, 131
Heath, Judge Richard, 148
Henchman, Bishop Humphrey, 25
Henley, 174
Henry VIII, King, ix, 95
Herbert, Arthur, 61, 140, 162
Herbert, Lord, 118
Herbert, Lord Chief Justice, 55, 57, 68
Hereford, 103, 105, 165, 166
Hervey, Lady, 82
Heywood, Oliver, 112
Hickes, George, 103
Hill, Christopher, 5
Hills, Henry, 92
Holdsworth, John, 22
Holland, 6, 12, 15, 31, 51, 52, 53, 56, 75, 76, 77, 80, 87, 121, 140, 180, 196
Holloway, Judge, 116, 130, 131, 148
Holt, Sir John, 114
Hooper, Bishop George, 31, 191, 192
Hornby, Charles, 80, 85, 141-2
Horneck, Anthony, 53
Hough, John, 63, 194
Huguenots, 11, 41, 46, 55, 102, 134, 188
Hull, 168, 171
Hulme, Lancashire, 60
Hungary, 102
Hungerford, 170
Huntingdon, Earl of, 57, 68, 146

'immortal seven', 14, 140, 142
invitation to William, 140-1, 160
Ipswich, 165
Isle of Wight, 28

James I, King, 80
James II, King (Duke of York), x, 1, 3, 4, 5, 6, 7, 8, 10, 11, 12, 16, 17, 18, 19, 20, 22, 28, 29, 31, 32, 34, 35, 36, 37, 38, 39, 40, 42, 43, 44, 45, 46, 47, 48, 51, 52, 53-9, 62, 63, 65, 70, 72, 73, 74, 75, 76, 77, 78, 80, 85-96, 98-113, 114-62, 163-93, 194-203

coronation, 37, 48-9, 50
Jeffreys, George, Lord Chancellor, 57, 62, 70, 99, 106, 107, 108, 111, 118, 121, 140, 200
Jervoise, Thomas, 162
Jesuits (Society of Jesus), 61, 71, 72, 78, 119, 120, 147
John, King, ix
Johnson, Samuel, 52
Johnstone, James, 75
Jolly, Thomas, 65
Jones, G. H., 42, 75, 91
Jones, J. R., x, 7, 9, 10, 16, 91
jury, 127, 132-9, 142, 147

Kemys, Sir Charles, 86, 133, 170
Ken, Bishop Thomas, 9, 21, 24, 28, 31, 41, 42, 43, 44, 46, 67, 85, 86, 89, 117, 152, 153, 154, 161, 164, 181, 189, 190, 191, 192, 193, 195
Kennett, Bishop White, 104
Kent, Earl of, 123
Kenyon, J. P., 10, 13, 14, 195
Kidder, Bishop Richard, 191
King's Bench, 106, 108, 115
'King's Evil', 41, 46, 67
kingship, sacramental view of, 23, 38, 39, 45, 160
Kirk, Colonel Percy, 142, 165, 167

Lake, Bishop John, 21, 23, 26, 29, 30, 39, 40, 45, 85, 88, 139, 143, 161, 181, 185, 190, 193, 195, 202
Lambeth, 84, 87, 90, 91, 109, 119, 133, 172, 180, 181, 182, 186, 189, 190
Lamplugh, Archbishop, Thomas, 85, 136, 157, 163, 172
Langley, Sir Roger, 124, 132
Lanier, Sir John, 165
Leeds, 66, 173
Leicestershire, 103
Levinze, Bishop Baptist, 119, 182
Levinze, Sir Creswell, 55, 114, 119, 125, 129
Leyburn, Bishop John, 76, 81, 145
Lichfield, 135, 136
Lichfield and Coventry, 30, 36, 45, 66, 103, 193
Lincoln, 30, 36, 45, 66, 103, 144
Lincoln's Inn, 71, 188

Lindsey, Lord, 162
Little Easton, Essex, 28
Llandaff, 32, 103
Lloyd, Bishop William (of Norwich), 85, 152, 181, 189, 190, 191
Lloyd, Bishop William (of St Asaph), 4, 6, 21, 24, 28, 32, 45, 46, 51, 80, 85, 88, 90, 91, 92, 106, 107, 117, 118, 121, 122, 139, 141, 142, 143, 150, 161, 164, 166, 174, 175, 180, 181, 183, 185, 186, 187, 192, 194, 195, 198
Lobb, Stephen, 66, 81
London, 25, 26, 48, 58, 59, 66, 67, 69–70, 71, 73, 75, 76, 80, 82, 83–96, 97, 100, 101, 109, 118, 133, 134, 140, 141, 147, 149, 150, 152, 153, 156, 158, 164, 165, 166, 168, 169, 170, 171, 172–5, 176, 186, 191, 199
  All Hallows, 26
  Ely House, 40
  fire of, 25
  Guildhall, 73
  Putney, 154
  St Botolph's Bishopsgate, 26
  St Giles Cripplegate, 84
  St Giles in the fields, 56, 59, 135
  St Martin in the fields, 32, 43, 71
  St Paul's Cathedral, 25, 26, 27, 29, 45, 56, 109, 189
  Savoy, 71–2
  Spitalfields, 188
  Tower of London, 1, 54, 109–13, 115, 117, 123, 134, 142, 143, 144, 158, 166, 172, 184, 185, 200
Longleat House, 191
Lonsdale, Lord, 105
lords lieutenant, 14, 16, 18, 43, 44, 46, 69, 70, 77, 123, 162, 166
Louis XIV, King, 7, 8, 11, 41, 52, 74, 102, 162, 166, 167, 175, 199
Lucas, Lord, 172
Lucock, Henry, 4, 5, 8
Lumley, Lord, 71, 124, 141, 163, 168
Lynn, All Hallows, 154

Macaulay, Lord, 3, 4, 5, 7, 13, 147
magistrates, 60, 69, 104, 158
Magna Carta, ix, 54

Maldon, 76
Manchester, 24
Manchester, Earl of, 123
martyrdom, 16, 23, 25, 45, 134, 169, 197
Marvel, Andrew, 27
Mary I, Queen, 3, 16, 135, 176, 197
Mary II, Queen (Princess of Orange), 4, 5, 6, 28, 30, 31, 53, 59, 71, 74, 78, 87, 90, 112, 114, 120, 121, 141, 176, 179, 180, 181, 182, 183, 184, 185, 186, 187, 188, 190, 191, 193, 202, 203
Mary Beatrice of Modena, Queen, 49, 64, 74, 87, 120, 123, 169, 171, 174, 181, 184
Massey, John, 62
Maynard, Sir John, 115
Mayo, Richard, 81
Merioneth, 28
metropolitan visitations, 30, 36, 37, 145
Mews, Bishop Peter, 51, 63, 85, 103
Middlesex, 124, 125, 135, 172
Middleton, Lady, 76
Middleton, Lord, 158, 166
Miller, John, 7, 10, 11, 12, 20
Mills, Henry, 165
Monmouth, Duke of, 10, 14, 33, 36, 39, 40, 44–4, 47–52, 63, 87, 88, 95, 106, 114, 149, 163, 164, 166, 170, 171, 198
Mordaunt, Lord, 174
Morley, Bishop George, 28, 34
Morrice, Roger, 65, 83, 84, 91, 136
Mulgrave, Earl of, 71
Mullett, Michael, 17, 192

Nantes, revocation of the Edict of, 8, 11, 52, 102
Naseby, Battle of, 63
Navy, Royal, 143, 153, 163, 172
Needham, William, 182
New York, 28
Newark, 26
Newcastle, 147, 164, 165, 168, 171
Newcastle, Duke of, 77
Newport, Lord, 124
nobility, role of, 13, 14, 46, 68–71, 76, 77, 116, 123–4, 146, 147, 171

Norfolk, 69, 152
Norfolk, Duke of, 59, 70, 71, 162
Northampton, Earl of, 53, 70
Northamptonshire, 58, 76, 135, 165, 171
Northumberland, 77
Norwich, 56, 58, 65, 100, 103, 135, 144, 147, 165
Nottingham, Earl of, 12, 14, 18, 48, 58, 61, 83, 86, 114, 120, 124, 127, 141, 159, 161, 169, 176
Nottinghamshire, 26, 77, 166, 167, 168

Oates, Titus, 46, 112
Oldham, Lancashire, 23, 24
Ormonde, Duke of, 140
Ossulston, Lord, 124
Owen, Sir Robert, 83
Oxford, 8, 21, 22, 23, 24, 26, 37, 51, 52, 59, 62–4, 65, 66, 67, 68, 76, 84, 97, 103, 119, 135, 140, 144, 145, 147, 148, 157, 165, 166, 169, 170, 178
  All Souls College, 168–9
  Christ Church, 28, 62, 136
  Corpus Christi College, 188
  Magdalen College, 15, 18, 62–4, 67, 68, 76, 103, 116, 136, 155, 156, 157, 194
  New College, 24
  University College, 62, 136
Oxford, Lord, 77

Papal Nuncio, 70, 71, 98, 156, 171
Parker, Bishop Samuel, 59, 63, 65, 66, 67, 76
Parliament, 10, 11, 12, 13, 22, 35, 36, 43, 50, 54, 61, 66, 68, 70, 77, 86, 87, 89, 101–2, 128, 129, 130, 148, 152, 155, 157, 158, 161, 162, 164, 166, 167, 168, 169, 170, 175, 195
  convention of 1688–9, 175–80, 189, 192
passive obedience, non-resistance, 12, 15, 82, 149, 178, 184, 194, 196
Patrick (Bishop) Symon, 56, 59, 79, 83, 84, 85, 97, 102, 135, 141, 150, 200

Pemberton, Sir Francis, 114, 125, 129
Pembroke, Earl of, 70, 123, 177
Penn, William, 81
Penyston, Charles, 68
Peterborough, 18, 30, 32, 39, 66, 84, 102, 136, 142, 154
petition of the bishops, 12, 85–96, 107, 125, 126, 157
Petre, Father Charles, 147
Petre, Father Edward, 54, 70, 76, 78, 90
Pincus, Steven, 17
Pinfold, Dr Thomas, 58
pluralities, 30
Plymouth, 166
politicians, 9, 18, 83, 147, 194, 198
Pollexfen, Henry, 114, 175
Poole, 144
Pope, 76, 79, 81, 156
Popham, Andrew, 60
Popham family, 170
Popish plot, 3, 27, 32, 46, 50, 51, 112, 135
popular rebellion, 13
Powell, Judge, 116, 131, 148
Powis, Thomas, 114, 115, 124
Powys, Lord, 55, 74, 76, 78
Preston, Lancashire, 79, 84, 173
Prideaux, Humphrey, 52
Prince of Wales ('Bedpan baby'), 1, 121–3, 136, 139, 169, 171, 181
Privy Council, 47, 52, 54, 55, 91, 106, 109, 122, 125, 126, 151, 158, 159, 162, 166, 191
propaganda, 11, 77, 92–6, 101, 119, 120, 134, 147, 165, 168, 197, 199
public opinion, 9, 14, 16, 17, 56
Pulton, Andrew, 71–2

Quakers, 56, 111
quo warranto proceedings against boroughs, 70, 75, 76, 77, 94, 144, 149, 152, 155, 156, 158, 163, 176

Radnor, Earl of, 124
Reformation, 93, 102, 108, 114
Reresby, Sir John, 70, 106, 125, 133, 136, 164

Index

Restoration of Charles II, 10, 20, 24, 25, 26, 27, 28, 44, 45, 63
Revelation, Book of, 45, 134
Richard II, King, 49, 99, 129, 169
Ripon, 28, 69
Rivers, Earl, 123
Rochester, 31, 171, 174, 186
Rochester, Earl of, 14, 32, 36, 57, 60, 61, 71, 83, 169, 172, 173, 175
Rome, 24, 74
Romsey, John, 86–7, 170
Rosewell, Thomas, 66
Rotheram, Judge, 148
Russell, Admiral, 141
Russell, Lord, 166
Rutland, Earl of, 70
Rye House plot, 47, 163
Ryle, J. R., 53, 111, 196, 197, 202

St Asaph, 18, 32, 51, 92, 142, 143, 154, 192
St David's, 66, 84, 144
St Winfriede's Well, 67
Sacheverell, Henry, 1, 2
Salisbury, 28, 36, 45, 135, 165, 167, 190, 191
Sancroft, Archbishop William, ix, 2, 6, 8, 9, 13, 15, 16, 21, 22, 23, 24, 25, 26, 27, 28, 29, 30, 31, 33, 34, 35, 36, 37, 38, 40, 41, 43, 44, 45, 46, 47, 48, 51, 53, 55, 57, 58, 59, 60, 68, 84, 85, 88, 90, 91, 103, 105, 106, 108, 109, 110, 114, 118, 122, 124, 132, 134, 137, 142, 145, 146, 147, 149, 150, 151, 153, 154, 155, 156, 158, 159, 160, 161, 164, 172, 173, 174, 176, 179, 180, 181, 182, 183, 186, 191, 192, 193, 195, 197, 202, 203
Savoy Conference, 25, 150
Sawyer, Sir Robert, 12, 106, 108, 114–39
Scarsdale, Earl of, 70, 123
Schomberg, Marshall, 162
Scotland, 27, 28, 31, 36, 44, 64, 72, 78, 135, 183, 196
Scott, John, 97, 150
Sedgemoor, 51, 63
Seymour, Sir Edward, 166

Shaftesbury, Earl of, 83
Sharp (Archbishop) John, 48, 53, 56, 57, 58, 79, 150, 178, 196
Sheldon, Archbishop Gilbert, 25, 26, 29
Sherlock, William, 16, 53, 56, 79, 83, 84, 100, 101, 149, 150, 196
Shrewsbury, 165, 171
Shrewsbury, Duke of (Earl of), 12, 61, 70, 71, 76, 120, 123, 141, 166, 174, 179
Sidney, Algernon, 116
Sidney, Henry, 52, 120, 123, 141, 142
Skelton, Colonel Bevil, 166
Slater, Samuel, 81
Smith, Bishop James, 81
Sodor and Man, 29, 30
Solomon, 39, 92
Somers, John, 114, 127, 129, 139, 140, 149
Somerset, 110, 150, 163, 168
Somerset, Duke of, 70, 143
Speck, W. A., 16, 131
Sprat, Bishop Thomas, 2, 54, 57, 58, 64, 66, 81, 97, 146, 160, 200, 201
Staffordshire, 69, 165, 171
Stamford, Earl of, 123
Stanley, William, 38, 53, 112, 114, 180
Stevens, John, 135, 143
Stillingfleet, Edward, 29, 42, 56, 61, 83, 84, 97
Stoke Newington, 118
Straka, Gerald, 5, 6, 195
Strickland, Admiral Roger, 143, 153
Strickland, Agnes, 4, 5, 90
Sudbury, 165
Sunderland, Earl of, 57, 58, 66, 69, 70, 74, 76, 87, 88, 90, 91, 106, 115, 118, 121, 125, 137, 152, 153, 164, 169, 195
Sussex, 77
Sussex, Earl of, 123

Tamworth, 135
Taunton, x, 163
Taylor, Bishop Jeremy, 94, 131

Tenison, Archbishop Thomas, 38, 51, 53, 56, 71–2, 79, 97, 112, 141, 150, 187, 189
Test Acts, 9, 11, 15, 43, 44, 46, 48, 50, 54, 55, 61, 64, 65, 69, 70, 73, 75, 78, 79, 93, 120, 135, 137, 146, 147, 152, 155, 158, 174, 198
Thanet, Earl of, 70
Therfield, Herts, 27
Thomas, Bishop William, 103, 143–4
three questions, 69, 76
Tillotson, Archbishop John, 56, 83, 84, 182, 193
tithes, 66
Treby, Sir George, 114, 158
Trelawny, Bishop Jonathan, 2, 3, 4, 16, 20, 21, 22, 28, 32, 33, 43, 44, 46, 68, 80, 81, 85, 86, 88, 92, 136, 137, 141, 142, 160, 161, 167, 177, 180, 181, 192, 195, 198
Trelawny, Colonel Charles, 142, 198
Trevelyan, G. M., ix, x, 44
trial of the bishops, 12, 19, 87, 114–39, 179, 198
Trinder, Serjeant, 114, 130
Tuckney, Anthony, 23
Turner, Bishop Francis, 9, 20, 21, 24, 27, 31, 44, 45, 46, 47, 49, 50, 85, 110, 132, 139, 142, 153, 154, 156, 161, 172, 175, 177, 181, 184, 185, 186, 187, 188, 189, 192, 195, 202
Twyford, 104
Tyrconnell, Lord, 55, 60

Uniformity, Act of, 53, 80
Utrecht, Treaty of, 193
Uxbridge, 164

Van Weede van Dijkvelt, Everard, 61
Vane, Sir Henry, 151
Vaughan of Carbery, Lord, 124
vicars apostolic, 76, 79, 81, 145, 155

Wagstaffe, Thomas, 189
Wake, Archbishop William, 56, 62, 67, 79, 80, 119, 150
Wales, 69, 83, 87, 97, 123, 131, 135, 153, 166
Walker, Obadiah, 62, 136, 148
Wallingford, 23
Walton, Isaac, 191
Ward, Bishop Seth, 85
Warrington, Lancashire, 173
Warwickshire, 69, 70, 136
Watson, Bishop Thomas, 84, 104, 144, 151, 171
Wells, 40, 41, 164, 165
Westminster, 24, 76, 115, 124, 172, 183
Westminster Hall, 55, 108, 116–39, 179
Weymouth, 166
Weymouth, Viscount, 191
Wharton, Henry, 60, 68, 166, 180, 182
Whig interpretation of history, ix, 3, 4
White, Bishop Thomas, 9, 21, 24, 26, 27, 30, 58, 64, 84, 85, 88, 89, 104, 106, 116, 125, 126, 139, 142, 161, 166, 181, 188, 189, 192, 195
Whitehall, 98, 105, 109–10, 133, 154, 174, 181
Wigan, Lancashire, 173
William I (The Conqueror), King, 177
William II (Rufus), King, 177
William III, King (Prince of Orange), 4, 5, 6, 7, 8, 12, 13, 14, 15, 18, 19, 28, 31, 32, 39, 51, 56, 60, 61, 67, 73, 74, 75, 76, 77, 78, 80, 90, 92, 114, 115, 122, 141, 142, 148, 152, 153, 154, 158, 159, 160, 161, 163, 164, 166, 167, 169, 170, 171, 174, 175, 176, 179, 180, 181, 182, 183, 185, 187, 188, 190, 191, 192, 199, 202, 203
Williams, Daniel, 81
Williams, William, 91, 98–100, 114–39, 148
Wiltshire, 169, 191
Winchelsea, Earl of, 171
Winchester, 63, 76, 135, 154, 193
Windebank, Sir Francis, 21
Windsor, 31, 67
Wolverhampton, 165, 171
Woodforde, Mary, 87

Worcester, 103, 143, 165, 170, 171, 193
Worcester, Marquis of, 123
Worcestershire, 69, 149
Wren, Christopher, 25
Wright, Lord Chief Justice, 98, 116, 117, 123–39
Wright, Nathan, Lord Keeper, 142

Yarmouth, 100
Yarmouth, Lord, 65
York, 25, 26, 29, 51, 76, 136, 147, 157, 164, 165, 168, 171
Yorkshire, 69, 140

Zadok, 39, 92